80

Advances in Biochemical Engineering/Biotechnology

Series Editor: T. Scheper

AF790240

Springer-Verlag Berlin Heidelberg GmbH

Process Integration in Biochemical Engineering

Volume Editors:
U. von Stockar · L. A. M. van der Wielen

With contributions by
A. Bruggink, J. M. S. Cabral, S.-O. Enfors, P. Fernandes, M. Jenne, K. Mauch, D. M. F. Prazeres, M. Reuss, S. Schmalzriedt, D. Stark, U. von Stockar, A. J. J. Straathof, L. A. M. van der Wielen

Springer

Advances in Biochemical Engineering/Biotechnology reviews actual trends in modern biotechnology. Its aim is to cover all aspects of this interdisciplinary technology where knowledge, methods and expertise are required for chemistry, biochemistry, microbiology, genetics, chemical engineering and computer science. Special volumes are dedicated to selected topics which focus on new biotechnological products and new processes for their synthesis and purification. They give the state-of-the-art of a topic in a comprehensive way thus being a valuable source for the next 3–5 years. It also discusses new discoveries and applications.

In general, special volumes are edited by well known guest editors. The series editor and publisher will however always be pleased to receive suggestions and supplementary information. Manuscripts are accepted in English.

In references Advances in Biochemical Engineering/Biotechnology is abbreviated as *Adv Biochem Engin/Biotechnol* as a journal.

Visit the ABE home page at
http://link.springer.de/series/abe/
http://link.Springer-ny.com/series/abe/

ISSN 0724-6145
ISBN 978-3-642-07818-7 ISBN 978-3-540-36782-6 (eBook)
DOI 10.1007/978-3-540-36782-6

Library of Congress Catalog Card Number 72-152360

http://www.springer.de

Originally published by Springer-Verlag Berlin Heidelberg New York in 2003
Softcover reprint of the hardcover 1st edition 2003

Typesetting: Fotosatz-Service Köhler GmbH, Würzburg
Cover: KünkelLopka GmbH, Heidelberg/design & production GmbH, Heidelberg

Printed on acid-free paper 02/3020mh – 5 4 3 2 1 0

Series Editor

Professor Dr. T. Scheper
Institute of Technical Chemistry
University of Hannover
Callinstraße 3
30167 Hannover, Germany
E-mail: scheper@iftc.uni-hannover.de

Volume Editors

Prof. Dr. Urs von Stockar
Laboratory of Chemical and Biochemical Engineering
Swiss Federal Institute of Technology (EPFL)
1015 Lausanne, Switzerland
E-mail: urs.vonstockar@epfl.ch

Prof. Dr. L.A.M. van der Wielen
Kluyver Laboratory for Biotechnology
Delft University of Technology
Julianalaan 67
2628 BC Delft, Netherlands
E-mail: luukvanderwielen@hotmail.com

Editorial Board

Advances in Biochemical Engineering/Biotechnology now also Available Electronically

For all customers with a standing order for Advances in Biochemical Engineering/Biotechnology we offer the electronic form via SpringerLink free of charge. Please contact your librarian who can receive a password for free access to the full articles. By registration at:

http://www.springer.de/series/abe/reg_form.htm

If you do not have a standard order you can nevertheless browse through the table of contents of the volumes and the abstracts of each article at:

http://link.springer.de/series/abe/
http://link.springer_ny.com/series/abe/

There you will find also information about the

- Editorial Board
- Aims and Scope
- Instructions for Authors

Editorial
Process Integration Challenges in Biotechnology Yesterday, Today and Tomorrow

1
Introduction

The industrial exploitation of biotechnology has proceeded through a number of distinct steps that were induced by scientific breakthroughs. After thousands of years of empirically based utilisation of microorganisms, the introduction of the science of microbiology in the mid nineteenth century created the opportunity to produce a number of chemicals by pure culture techniques. These products were mainly limited to organic acids and alcohols due to the problems of running large scale submerged cultures under aseptic conditions. The next breakthrough was made during the development of the penicillin process during the 1940s, which was the result of a concerted action on the integration of classic genetics, organic chemistry and chemical engineering. This integration of engineering and biosciences led to the emergence of the *biochemical engineering* discipline. The bioprocess technique that was then created formed the basis for a large number of industrial processes for the production of products based on microbial metabolism, such as antibiotics, enzymes, amino acids, vitamins etc. However, the technique was restricted to the use of the organism in which the exploited gene/metabolic pathway was found in Nature.

The third biotechnical breakthrough in the 1970s, was based on the developments in molecular genetics that were first adopted for the production of heterologous proteins in microorganisms and animal cell cultures. This scientific breakthrough extended the application potential of biotechnology by a quantum leap. Some of the immediate outcomes concerned the production of highly valuable proteins especially for medical and analytical purposes, which hitherto could only be extracted from whole organisms or were unavailable altogether. However, the impact on bioprocessing was equally far reaching in that the biocatalytic activity and the host organism could now be decoupled. While the production was previously limited to the use of the species in which the gene of interest was found, the gene is now a source of information that can be inserted into hosts that are best suited to industrial production, such as *E. coli*, *Bacillus spp.*, *Aspergillus spp.*, yeasts, CHO and insect cells. The ever-increasing know-how concerning the handling of genes and their transfer from one organism into another gave rise to the possibility of considering production of a given product in a stunning variety of living systems including procaryotic and eucaryotic microbes, cell cultures, eggs, transgenic plants and animals.

While bioprocessing was recognized as a highly elegant and specific way to produce extraordinarily complex molecules under mild reaction conditions, it was also perceived as an inherently low productivity production system relative to chemical processes, which results in voluminous process equipment. This low productivity is mainly caused by the fact that biocatalysts such as cells and enzymes have evolved in nature to function optimally in a low concentration environment. This is the reason why biotechnology is often so much superior to chemical technology in environmental applications, while suffering from inhibition problems when engineers try to use them in concentrated environments. Other biocatalytic agents, such as animal cells, are intrinsically able to build up very high cell densities in their natural environments, but grow to only very low cell numbers in bioreactors, basically because their extremely complicated nutritional and culture condition demands are not understood well enough. Process productivity often also suffers from degradation of the products in the reactor or during the downstream processing. Another inherent problem is the high degree of purification that is required for some of the (pharmaceutical) bioproducts. This requires a multi-step downstream processing with an inevitably low overall product yield.

As the impact of choices made in the initial stages of a bioprocess (upstream processing) is perceived in later stages (bioreactor, downstream processing), any improvement of the situation and the development of more efficient bioprocesses relies strongly on the balanced interaction of rather different disciplines from the technical sciences and the biosciences. However, until the nineties no international research programme had ever addressed this field. This has meant that the important linkage between the fundamental developments in the biosciences and the possible industrial applications was completely missing.

2
ESF Programme Process Integration in Biotechnology (PIBE)

Following similar considerations, a working group for Technical Science of the European Science Foundation (ESF) has identified in 1990 'process integration in biotechnology' as being of high priority in that it links basic technical sciences to the fundamental biosciences. Based on the results of a Workshop on Process Integration held on 7–8 December 1990 in Frankfurt-am-Main, Germany, a proposal for an ESF Programme on Process Integration has been prepared by its chairman, Professor Karel Luyben of the Delft University of Technology in the Netherlands. It was presented at the April 1991 annual meeting of the ESRC and received strong support. At its September 1991 meeting, the ESF Executive Council recommended the Programme for launching by the 1991 General Assembly for a period of three years. In 1991, the General Assembly launched the ESF Programme on Process Integration in Biochemical Engineering. The ESF Programme aimed at enhancing the interdisciplinary approach towards integrated bioprocessing that includes protein, genetic, metabolic and process engineering to link basic developments in the biosciences with possible industrial applications. The purpose of the ESF Programme was to establish a platform for strong European research groups in this field to strengthen and to stimulate the input of Bioprocess Technology (Biochemical Engineering), which could bridge the gap between basic biosciences and process development.

The programme on Process Integration in Biochemical Engineering, comprised different lines that will be characterized briefly.

2.1 Workshops

A series of workshops was organised at the frequency of 1–2 workshops per year. The goal of these workshops was to present and to elaborate current approaches around a particular theme in the PIBE field and to generate new ideas for collaborative programmes of research between laboratories. The emphasis is on bringing together younger scientists and a smaller number of senior scientists, chosen with reference to their expertise.

The topics of the workshops were 'Integrated Downstream Processing' (Delft, the Netherlands, 1993), 'Integrated Upstream Processing' (Sitges, Spain, 1993), 'Intensification of Biotechnological Processes' (Davos, Switzerland, 1994), 'Integrated Environmental Bioprocess Design' (Obernai, France, 1995) and 'Integrated Bioprocess Design' (Espoo, Finland, 1996). The number of participants for each workshop was typically restricted to 40, and equally distributed over senior and junior scientists. The outcome of each individual workshop was summarized in a workshop report.

2.2 Short-Term Visits

Exchange of younger scientists working for their PhD as well as senior scientists for shorter period of time is extremely beneficial for fast and efficient exchange of information and ideas. In view of the multidisciplinarity of the field of biochemical engineering, stimulating these exchanges was an important aspect of the PIBE programme. However, to elaborate a certain part of a project within an interdisciplinary project or to initiate a common international research programme, transfers in the order of 2–4 months were necessary and desirable.

2.3 Graduate Course on Thermodynamics in Biochemical Engineering

Rational and efficient process development in chemistry always makes heavy use of thermodynamic analysis. It is evident that biotechnologists have shunned this field for whatever reasons. The Steering Committee of the PIBE programme concluded that this state of affairs was one of several reasons why development and design of biotechnological processes is today mostly carried out in an essentially empirical fashion and why bioprocesses often are not as thoroughly optimised as many chemical processes. It therefore decided that for efficient process integration it was necessary to stimulate a more systematic use of thermodynamics in the area. Recognizing that quite a large body of knowledge in the area of biothermodynamics already existed, it was decided to develop a course for advanced graduate students and researchers to make the field of applied thermodynamics in biotechnology better known and to stimulate its use. Meanwhile, this graduate course on *Thermodynamics in Biochemical Engineering* has taken place four times: 1994 in Toulouse (France), 1996 in Braga (Portugal), 1998 in Nijmegen (The Netherlands) and 2000 on Monte Verità above Ascona (Switzerland).

2.4 Platform

By integrating the results from the two points above, it was possible to establish the Section of Biochemical Engineering Science within the European Federation for Biotechnology as a sustainable entity. The Section of Biochemical Engineering Science is meant to be a platform within the field of Bioprocess Technology, aimed at promoting this field and contacting academics and industrialists by organising conferences and other activities, as well as to advise the direction and focus of the research programme of the EC.

2.5 Conclusion

After the end of the 1990s during which the ESF Programme on Process Integration in Biochemical Engineering was conducted, it was appropriate to look back on this work and try to assess what had been achieved. The following series of articles have been written by scientists and engineers who have made important contributions to the programme. They report some of the major findings, limits and challenges of bioprocess integration.

3 Future Challenges in Process Integration in Biotechnology

Today, biotechnology is accelerated by rapid scientific developments in molecular biology, protein chemistry and information technology, which push the sciences of microbial and cell physiology forward at a high speed. Thus, a number of bioengineering tools are currently discussed, investigated, and exploited, each building on an integration of previous tools with new scientific knowledge and techniques (Table 1).

Table 1. Engineering tools resulting from the integration of different scientific areas

Scientific Basis	"Engineering" Tool	Application
Molecular genetics	Genetic engineering	Production of heterologous proteins
Protein chemistry	Protein engineering	Production of improved or novel proteins
Metabolism	Metabolic engineering	Production of metabolites
Physiology	Physiological engineering	Design of improved host cells
Medical and material sciences	Organ engineering	Design of artificial organs

The current task of biochemical engineering research and development is to integrate and develop the new tools for the industrial applications. The borders between the traditional activities in bioprocessing, often called upstream, reaction and downstream processing, respectively, are becoming more and more diffuse due to these developments. Each of the listed "engineering" tools may play a role in each of these traditional activities in the exploitation of the cells/biomolecules:

Protein engineering is used for the design of protein products with improved properties, or with altogether novel functionalities, for bioprocessing, the design of new separation and for analytical methods. Although proteins are the basic molecular machines that we exploit in biotechnology, our understanding of their function and how this depends on structure is still very incomplete. Enormous challenges lay ahead. Protein chemistry must be integrated with classical physical chemistry and chemical engineering tools dealing with biothermodynamics, adsorption/desorption kinetics, mass transport and modelling.

Metabolic engineering was first considered to become an easy application of the genetic engineering tool. However, the relatively few successful applications so far, for example the production of aromatic amino acids with *E. coli*, and the numerous as yet less successful efforts to eliminate the overflow metabolism of glucose by *E. coli* and *S. cerevisiae*, show that this approach, albeit realisable, needs a much deeper understanding of the regulation of the metabolism. To achieve this, extensive work on metabolic flux analysis and modelling must be combined with the genetic engineering tool. Once again, the advanced modelling needed for this will demand an integration of not only metabolism and analytical chemistry, but also of high-performance reactor design, advanced rapid on-line monitoring and new methods for the mathematical modelling of the control of complex systems.

Physiological engineering widens the concept to controlling/designing the cell with other properties that are important for its application, such as membrane, cell surface and organelle properties, resistance factors and protein processing functions. In this way, hosts with more process-fitted properties will be designed. The tools are there, but the target must be selected based on an understanding of the cell-environment interactions.

Improvement of cells and/or process control strategies must be based on a deeper understanding of the function of the cell under process conditions. It means a demand for research on the cell-environment interactions. This is a well-established research field in environmental microbiology, where the time-frame is usually hours or days, but the analysis of for example physiological stress responses and *corum sensing* and transcriptional control is also needed with the time-frame of seconds under process conditions in order to better understand the organism and to design the control or the cell for the process. Taken together, these techniques provide the tools for *biosystems engineering*.

Organ engineering requires an equally challenging integration of molecular biology, protein chemistry, physical chemistry of surfaces, and medical and material sciences. The design of artificial organs shows similarities with the design of a bioreactor for production purposes, and will therefore also require the integration of all these disciplines with biochemical engineering.

New targets for biochemical engineering. Most of the discussion above, and the applications of biochemical engineering so far have been limited to industrial production purposes. However, the biochemical engineering science will also play a major role in new applications in which large numbers of different cells or enzymes are handled, characterized, selected, and utilized under precisely controlled reaction conditions. The developments in functional genomics, proteomics and high-throughput screening for drug development put an increasing demand on rapid reproducible production of proteins for analytical purposes. A similar demand exists for the rapid characterization of recombinant production strains and other industrial biocatalysts. Contrary to the traditional bioprocessing, satisfying such demands needs the development of smaller and smaller reactor volumes equipped with the same potential for rapid on-line analysis, modelling and reproducible process control as the high-performance laboratory reactors of today. This development may ultimately lead to controlled cell micro-bioreactors and nano-enzyme reactors. Furthermore, these might be integrated with the currently developed analytical nanosystems (the "lab-on-a chip" concept). Thus we will witness a certain coalescence and integration between the fields of functional genomics, transcriptomics, proteomics, metabolomics and biochemical engineering.

4 Conclusions

Bioprocess integration has been shown to be one of the key prerequisites for improving the efficiency of industrial biotechnology and for transforming bioprocess and bioproduct technology into a science-based, rational engineering discipline. However, a short qualitative analysis of possible future trends in biotechnology and biochemical engineering will require the coalescence of even more, widely different scientific disciplines. The success of these foreseeable trends will amongst other things depend on how well these disciplines can be integrated. Despite the fact that being highly proficient in any given field of science and engineering requires a good deal of specialisation, sufficient attention must be given to the integration of different disciplines. International efforts such as the ESF programme on bioprocess integration could undoubtedly make powerful contributions in this respect.

October 2002

Sven-Olaf Enfors
Luuk van der Wielen
Urs von Stockar

Contents

Back to Basics: Thermodynamics in Biochemical Engineering

U. von Stockar[1] · L. A. M. van der Wielen[2]

[1] Institut de Génie Chimique, Swiss Federal Institute of Technology, 1015 Lausanne, Switzerland. *E-mail: urs.vonstockar@epfl.ch*

[2] Kluyver Laboratory for Biotechnology, Delft University of Technology, 2628 BC Delft, The Netherlands. *E-mail: luukvanderwielen@hotmail.com*

Rational and efficient process development in chemical technology always makes heavy use of process analysis in terms of balances, kinetics, and thermodynamics. While the first two of these concepts have been extensively used in biotechnology, it appears that thermodynamics has received relatively little attention from biotechnologists. This state of affairs is one among several reasons why development and design of biotechnological processes is today mostly carried out in an essentially empirical fashion and why bioprocesses are often not as thoroughly optimized as many chemical processes. Since quite a large body of knowledge in the area of bio thermodynamics already existed in the early nineties, the Steering Committee of a European Science Foundation program on Process Integration in Biochemical Engineering identified a need to stimulate a more systematic use of thermodynamics in the area. To this effect, a bianual course for advanced graduate students and researchers was developed. The present contribution uses the course structure to provide an outline of the area and to characterize very briefly the achievements, the challenges, and the research needs in the various sub-topics.

Keywords. Thermodynamics, Phase equilibria, Biotechnology, Biochemical engineering, Biomolecules, Irreversible thermodynamics, Energy dissipation, Living systems

Advances in Biochemical Engineering/
Biotechnology, Vol. 80
Series Editor: T. Scheper

1 Introduction

Most quantitative theories and calculations in engineering sciences rely on a combination of three fundamental concepts: balances (e.g., mass, energy, elemental, momentum), equilibria (e.g., force, reaction, phase equilibria), and kinetics (e.g., momentum, mass and heat transfer, enzymatic and growth kinetics). While balances and kinetic models are used extensively by biotechnologists, the same is not true for thermodynamics, and the equilibrium aspects and non-equilibrium thermodynamics appear to be largely disregarded by many of them.

In the early nineties, the Steering Committee of the European Science Foundation (ESF) program on Process Integration in Biochemical Engineering (PIBE) therefore decided that for efficient process integration it was necessary to stimulate a more systematic use of thermodynamics in the area. Since quite a large body of knowledge in the area of biothermodynamics already existed, it was decided to develop a course for advanced graduate students and researchers to make the field of thermodynamics as applied to biotechnology better known and to stimulate its use [1]. The authors of this article were given the task of organizing and coordinating the events. Meanwhile, this graduate course on Thermodynamics in Biochemical Engineering has taken place four times: 1994 in Toulouse (France), 1996 in Braga (Portugal), 1998 in Nijmegen (The Netherlands), and 2000 on Monte Verità above Ascona (Switzerland). The contents of the more recent editions of the course as well as the lecturers are summarized in Table 1.

The present review uses the structure provided by this course to give a very short outline of the field and to present some brief remarks concerning the state of each topic. This is an update of a similar review that appeared some years ago [2].

Process integration in biochemical engineering depends on the application of thermodynamics because for rational development and optimization of processes engineers need ways and means to estimate biomolecular properties, thermodynamic equilibrium positions, driving forces, energy efficiencies and the like. The importance of thermodynamics in obtaining such data is summarized in Table 2. The relative scarcity of pertinent data of this kind and the failure to use thermodynamic tools to estimate them, is one among several reasons why development and design of biotechnological processes is today mostly carried out in an essentially empirical fashion and why bioprocesses are often not as thoroughly optimized as many chemical processes.

Rigorous application of thermodynamics to bioprocesses may seem a daunting task in view of the astronomical complexity of the reaction mixtures, giant biological molecules, intramolecular forces, multiple driving forces, and the multitude of phases and biological, chemical, and physical processes which have to be dealt with. However, rational, efficient, and rapid process development and equipment design can only be achieved on the basis of a sound scientific foundation, as it is available nowadays, for example, for the petrochemical industries [3]. The more extensive use of thermodynamics and

Table 1. Contents of the course on thermodynamics for biochemical engineers

Course subjects	Potential applications
Fundamentals	
– *Phase Equilibrium Thermodynamics of Non-Electrolytes*, J.M. Prausnitz. Homogeneous mixtures, excess properties, VLE, SLE, LLE in non-electrolyte systems, activity coefficient models.	General insight into equilibria
– *Obtaining thermodynamics properties*, C.A. Haynes. Direct methods and the use of Gibbs-Duhem equation	
Large and charged species	
– *Electrolytes*, C.A. Haynes.	Solution behavior of polymers and proteins, salting out, precipitation, extraction, chromatography, resin swelling, phase splitting etc. General relevance for DSP
– *Polymers, polyelectrolytes, gels, demixing in polymer solutions, Donnan effect, swelling in hydrogels*, J.M. Prausnitz	
– *Aqueous two-phase systems*, C.A. Haynes.	
– *Correlative approach for complex biomolecules*, L.A.M. van der Wielen.	
– *Phase equilibria in protein solutions*, J.M. Prausnitz. Integral theory of solution, potentials of mean force, RPA theory	
Proteins and biocatalysis	
– *Conformational and structural stability of proteins*, W. Norde. Enthalpic and entropic effects, salts, solvent, temperature, denaturation, renaturation	Biocatalysis in general and in non-conventional media biocatalyst engineering, protein engineering, DSP, inclusion body reprocessing
– *Phase and reaction equilibria in biocatalysis*, P.J. Halling (in 1994). Effects of cosolvents, pH, and salts	
Irreversible thermodynamics	
– *Thermodynamics of open and irreversible systems*, U. von Stockar.	Insight, coupled fluxes in cellular and process scale membrane processes, ion exchange, and living systems
– *Mass transfer on the basis of IT*, L.A.M. van der Wielen. Multicomponent diffusion, multiple driving forces, flux coupling	
Thermodynamics in living systems	
– *Energy dissipation in biotechnology*, U. von Stockar. Heat generation, free energy dissipation, and growth. Energy balances, biocalorimetry, and monitoring of bioprocesses	Insight, heat removal, monitoring of bioprocesses, prediction of biomass and product yields, metabolic engineering
– *Description of microbial growth based on Gibbs energy*, J.J. Heijnen. Yield and maintenance correlations	
– *Opening the black box: thermodynamic analysis of metabolic networks*, U. von Stockar, C. Cannizzaro. Genomics, metabolomics and metabolic flux analysis, thermodynamic feasibility, computer demonstration	Metabolic pathway feasibility analysis based on thermodynamics

Table 2. Potential role of thermodynamics in biotechnology

- Prediction of physical-chemical properties of biomolecules
- Prediction of phase equilibria, in particular for DSP, and reaction equilibria, in particular for biotransformation
- Structural and functional stability of proteins and other biomolecules
- The effect of *T*, pH, *P*, solvents, and solutes on activity and selectivity of biocatalysts
- Correct formulation of driving forces for bioprocesses
- Thermodynamic effects in cellular growth, including heat generation
- Efficiency of cellular metabolism: Optimal biomass and products yields
- Quantification and improvement of the efficiency of bioprocesses with respect to the use of raw materials, auxiliary materials, and energy

especially its further development for the complex world of biochemical engineering therefore remains one of the major challenges in biochemical engineering.

2 Phase Equilibria of Large and Charged Species

Benzyl penicillin (penicillin G) is one of the smaller biomolecules of industrial relevance, which is already fairly large when compared to many petrochemicals. Biomolecules are a large group of polymers and most bear pH-dependent charges. This is one reason why the excellent predictive models available today for non-charged, small chemicals, cannot be used straightforwardly in biochemical engineering. A characteristic example is the description of the phase behavior of penicillin G in water-alkylacetate esters, which are typical industrial solvent extraction systems. Despite its industrial scale of operation (estimated as 10^4 t year^{-1} in 2000) and its 50-year history, phase equilibria have hardly been dealt with in great detail. Using one of the more powerful predictive models (UNIFAC), partition coefficients over organic and aqueous phase are overestimated by several orders of magnitude. Even worse, tendencies for homologous series of solvents are predicted completely erroneously, as shown in Fig. 1. This implies that design and optimization for these and even more complex processes have to follow the laborious and costly empirical route, rather than use computer-aided flowsheeting programmes for the evaluation of alternatives. This is an area in which molecular thermodynamics can make a useful contribution [4].

Therefore, the cluster of topics around the phase behavior of large molecules and charged species is one of the absolutely central themes in biothermodynamics. It forms an essential basis for instance, for all possible forms of bioseparation processes (Table 2). In some of these areas, a huge body of research is currently active. Basically three approaches can be distinguished. These are (1) the extension of existing methods and excess models (NRTL, UNIQUAC etc.) to aqueous, electrolyte systems containing biomolecules [5, 6], (2) osmotic virial and closely related models based on the consideration of attractive and repulsive interactions between solutes via potentials of

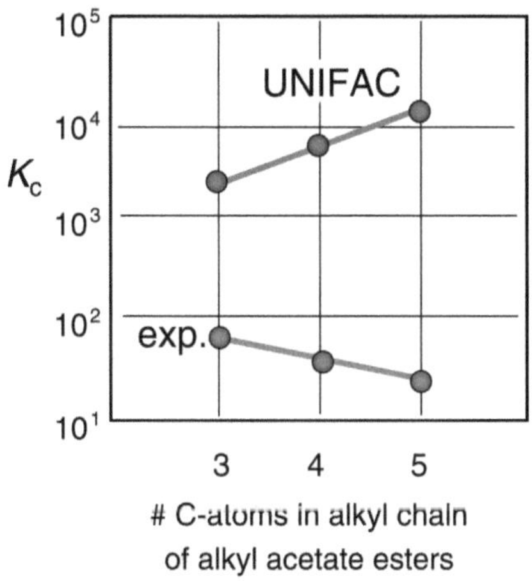

Fig. 1. Experimental partition coefficients of penicillin G (K_{PenG}) and those predicted using UNIFAC

mean force [7], and (3) correlative methods based on rigorous thermodynamics [8, 9].

The development of experimental tools to obtain the essential parameters from independent data, and the development of estimation techniques for these parameters are crucial in this field. Among the former, laser scattering methods (mainly for macromolecules), membrane osmometry [4], and potentiometric methods [10] should be mentioned. A challenging example of the impact of the increased availability of these methods is the large-scale crystallization of proteins. Protein crystallization has always been notoriously difficult to predict. It has been shown by George and Wilson [11], that the production of pure protein crystals, instead of amorphous and contaminated precipitates, is possible only in a narrow 'window of operation'. This region is determined relatively easily using the abovementioned methods.

Quantitative, correlative approaches based on hydrophobicity, polarity, and the Hansch parameter have proved to be useful and consistent in aqueous two-phase extraction [12], reversed micellar extraction [13], reversed-phase, hydrophobic interaction [14], and ion exchange chromatography [15, 16], as well as solubility in mixed solvents [8, 9, 17].

Figure 2 gives an example of the potential of correlative methods. The curve, calculated with a relatively simple correlative method of [8, 9], should be compared to the straightforward extension of conventional, Born theory-based models (area) for the solubility of the amino acid L-valine in an alcohol-water mixture (markers).

However, thermodynamic considerations in areas such as protein fractionation by precipitation, chromatography, solvent extraction, aqueous two-phase systems, and the like in order to understand the partitioning and other effects at least qualitatively are still underdeveloped and should receive increasing attention [14, 18–21].

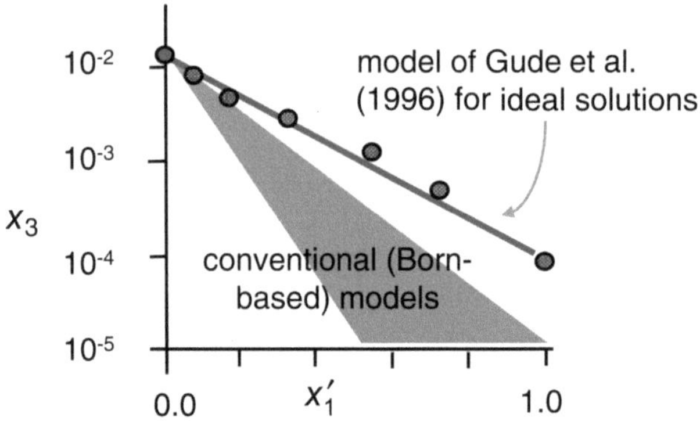

Fig. 2. Experimental (*grew circles*) and predicted (*curves, area*) solubilities of valine (*3*) in water (*2*) + ethanol (*1*) mixtures as a function of the solute-free mole fraction ethanol (x_1')

Another field of increasing interest in biotechnology related industries is that of heterogeneous structures: colloids, micelles, bilayer membranes, foams, and (hydro)gels. Living systems are composed largely of polymers (polysaccharides, proteins), which possess colloidal properties by virtue of their size, but which can self-assemble into a great variety of organized structures [22]. Technical applications can be found, amongst others, in food and feed, drug formulation and delivery in pharmaceutics, consumer products, technical foams, paints, chromatographic resins, and superadsorbing materials. The role of electrostatic and hydrophobic effects and their interaction on colloidal phenomena can nowadays at least be described qualitatively and, increasingly, quantitatively.

Swelling equilibria of charged and uncharged (hydro)gels can be described with a combination of Flory-Huggins theory, elastic deformation, and electrostatic effects [4]. A typical example is ion exchange chromatography of weak electrolytes (proteins in buffered solutions), where chromatograms can only be interpreted quantitatively when solute partitioning is described using above elements [23–26] as demonstrated schematically in Fig. 3. It has also been shown that the equations describing the swelling equilibria provide an excellent basis for the description of the dynamics of the swelling process itself [27]. This includes the description of the internal structure development of the swelling gel.

Literature on thermodynamics of biopolymers other than proteins, such as DNA, does not seem to be available in large amounts. It is conceivable that this area might become important due to the fact that the scale at which DNA will have to be isolated and purified will become considerably larger in the future, as such areas as somatic gene therapy, DNA immunization and vaccination, and transient expression of gene products for rapid production of preparative amounts of recombinant proteins gain wider interest [28–31].

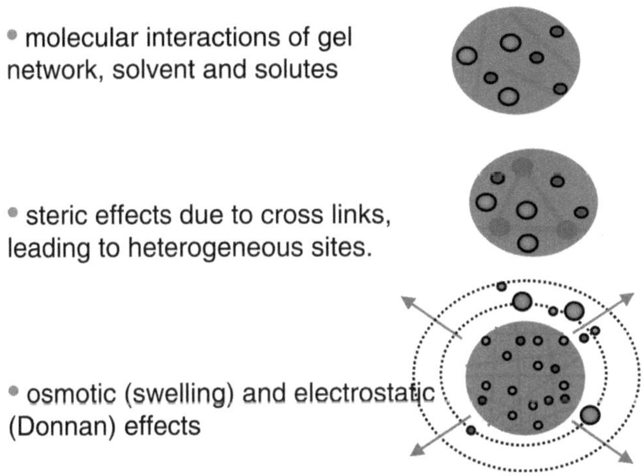

Fig. 3. Effects affecting partitioning of relatively large biomolecules over liquid and resin phases

3
Proteins and Biocatalysis

Another major area of impact of thermodynamics concerns the structural and functional stability as well as the activity of the proteins. The technical implications of knowledge in this field for reprocessing recombinant proteins by unfolding and refolding and for designing appropriate micro-environments and processing conditions in bioreactors and recovery equipment are evident. The lectures on conformational and structural stability of proteins are thus a key element in the course.

It is probably less appreciated that thermodynamics is also of great importance in understanding protein function. This was recognized many years ago by the EFB Working Party on Applied Biocatalysis, who in 1992 organized an international symposium on Fundamentals of Biocatalysis in Non-Conventional Media to stimulate the development of a clear scientific base for biocatalysis using non-aqueous solvents [32, 33].

Thermodynamic effects on biocatalysts working in the presence of non-conventional media have an impact on two levels: i) phase and reaction equilibria and ii) biocatalyst stability and activity [34]. The thermodynamic effects on the first level are by now relatively well understood. It is probably safe to say that a certain scientific foundation for rational "phase and reaction equilibrium engineering" exists. Based on this knowledge, it is possible to conceive, if not to design, biocatalytic systems with tailored selectivities and/or improved product yields due to low water activity, the presence of non-aqueous non-conventional solvents [33], or characterized by a very high solid content [35, 36]. It has been shown for particular cases that this type of engineering may be based directly on standard thermodynamic tools such as UNIFAC calculations [37]. Nevertheless, much work remains to be done in this area.

The situation is worse on the level of the biocatalytic molecule itself (ii). Solvent molecules, residual water molecules in low-water environments, temperature and pH all affect the stability, activity, equilibrium conversion, and product distribution in a variety of ways, some of which, as for example, the influence on the free energy of the substrate in the ground and the transition states, must be analyzed in thermodynamic terms. Even if our qualitative understanding of such effects is improving, we are still far from a complete description, which will require much more thermodynamic work in this area.

One of the most pretentious approaches for future biochemical engineering would consist of tailoring proteins to desired functions by protein engineering. Pioneering work has for example been done in the area of biocatalysis, but it is commonplace that rational exploitation of protein engineering will require an enormous amount of additional knowledge on the primary – tertiary structure – function relationships. These again emphasize the importance of thermodynamics in the area of protein stability.

4 Irreversible Thermodynamics

4.1 Multicomponent Transport

Another characteristic of living and technological systems is the frequent occurrence of multiple fluxes and flux coupling at various levels at various scales of scrutiny. Although it is possible to describe mass transfer effects based on Fick's law-type equations [38], the solutions may become involved and awkward. This is why the 'novel' and much more elegant approach based on ideas [39, 40] and irreversible thermodynamics and elaborated by Wesselingh and Krishna [41] is introduced. The resulting rate equations are, however, completely unfamiliar to most engineers and their use must be stimulated by advanced courses such as the present one. The same approach is in principle possible for obtaining other transport properties such as the viscosity of water-cosolvent mixtures when compared to water. This is illustrated in Fig. 4, in which calculated classical Fickian diffusivities and viscosities of ethanol-water mixtures using the Van Laar model are compared to the respective experimental data. Calculated curves are for ideal systems (linear: logarithmic interpolation) and real systems (curves). Viscosity data in Fig. 4 are from Wei and Rowley [42]. The ideal diffusivity has been calculated using the Vignes [43] approximation, whereas the real curve for the predicting Fick's diffusion coefficient is based on the Stefan-Maxwell diffusivities combined with the Van Laar equation for estimating the activity coefficients.

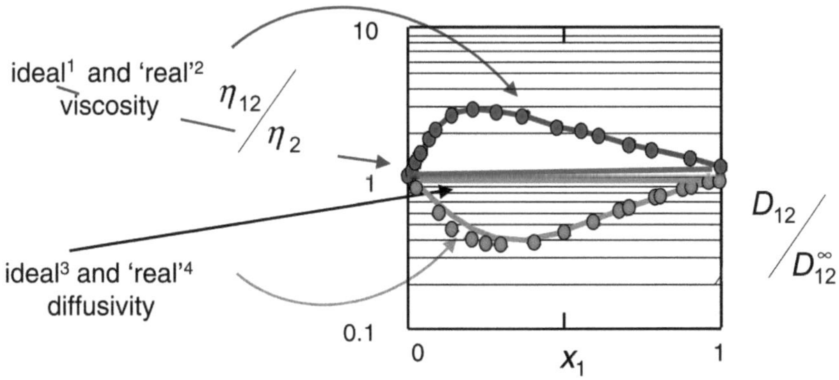

Fig. 4. Relation between ideal and real viscosities (*upper curve* and *markers*) and diffusivities (*lower curve* and *markers*) and composition in an ethanol (*1*) + water (*2*) system [1–3]; [4] using Van Laar model

4.2 Exergy Analysis and Efficiency of Processes

The Second Law of Thermodynamics tells us that all real processes inevitably lead to entropy production or, formulated differently, to a lower energetic quality of the product flows compared to the input flows [44]. The energetic quality of a process stream is expressed in terms of exergy [45], which quantifies the (remaining) Gibbs free energy that *can* still be extracted from the system. In *real* biotechnological processes, pure or highly concentrated materials such as sugars and salts are mixed at great exergy loss in huge quantities of water to produce relatively pure but otherwise useless gaseous CO_2 and very dilute product streams. The problems created here have to be solved in the downstream processing train.

The recovery and purification of the desired product demands a further breakdown of exergy in the sense of '*mixing*' the aqueous feed with (pure) solvents (precipitation and extraction), salts (ion exchange), heat (evaporation and solvent recovery), electrical power (electrodialysis), pressure (filtration and membrane separations), or just extra water (gel filtration). This is shown schematically in Fig. 5.

Useful work is usually proportional to flux (N) of a species through the process, and hence is more-or-less proportional to its driving force (in Fig. 6 given as a chemical potential gradient). Lost work is given by the product of flux (N) and driving force, and is therefore proportional to the squared driving force. At low driving force, only small amounts of work are lost, but also the capacity of the process is low, which is undesired. At high driving forces, however, lost work (proportional to squared driving force) may well exceed useful work. Operation at intermediate driving force appears attractive to optimize the ratio of useful and lost work. This is demonstrated in Fig. 6.

Probably the most beautiful feature of exergy is the unified description of the quality loss of energy and (auxiliary) material streams in terms of kJ mol^{-1}. This

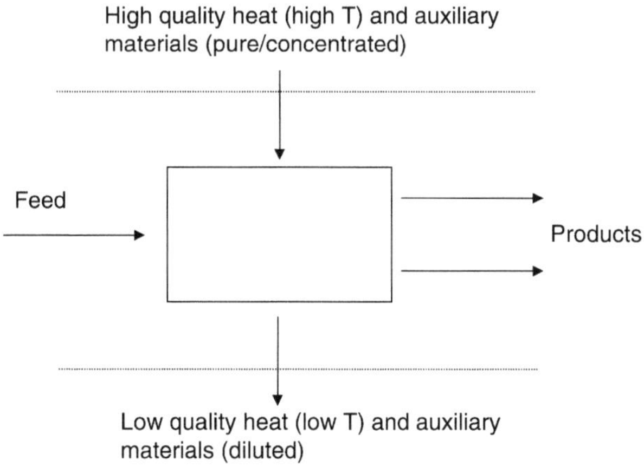

Fig. 5. Processes as open systems, driven by the input of heat and auxiliary materials

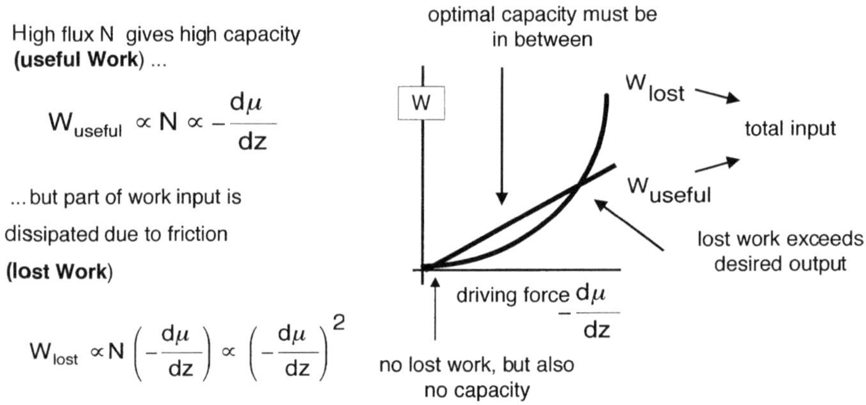

Fig. 6. Relation between useful work, lost work and magnitude of driving force (and process capacity)

provides a unified basis for comparison of fairly different process set-ups. This is not possible with other indices for process quality such as heat consumption or the *EQ*-factor (kg waste per kg of product) of Sheldon [46].

An example is the recovery and purification of amino acids via crystallization. Here, the solubility of the amino acid can be influenced by a number of methods: (1) lowering the temperature, (2) evaporating the solvent, (3) selective removal of the solvent by means of membranes techniques, and (4) by using a water-miscible cosolvent such as lower alcohols and acetone. In the last of these, which is close to industrial practice, work is lost at a large number of places. Unequal 'quality' of heat input (at a high *T* level) and recovery (at a low *T* level) and incomplete solvent recovery from the mother liquor increase lost work and, less obviously, incomplete recovery contributes to lost work as well. This is shown schematically in Fig. 7. Considering option 3, work is lost to force the solvent (wa-

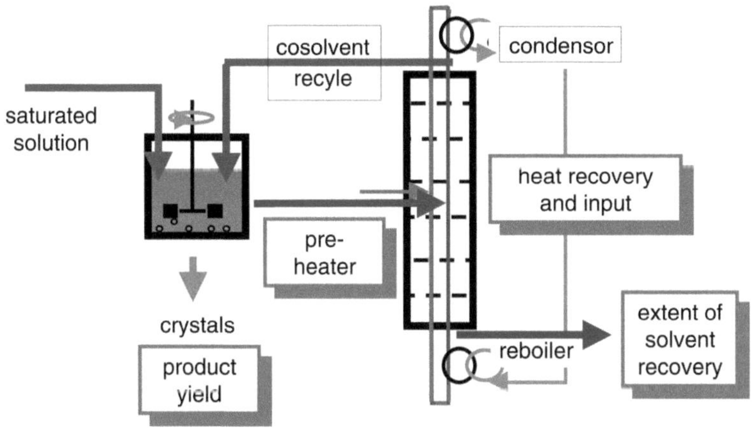

a

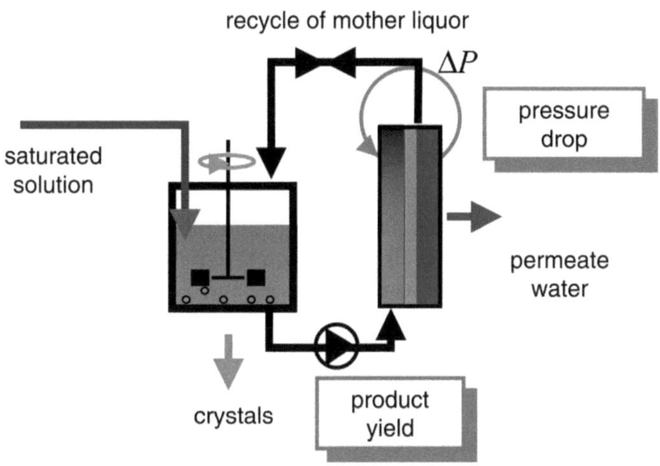

b

Fig. 7. **a** Locations for the large losses of exergy in crystallization of amino acids using a water-miscible cosolvent (*shaded boxes*). **b** Locations for the small losses of exergy in crystallization of amino acids using a selective (e.g., nano-filtration) membrane

ter) flow through the membrane at a more-or-less constant pressure drop and, less obviously, in the form of incomplete recovery. Obviously the exergy loss of both configurations is not equal, and can be quantified during flowsheeting. Therefore, analysis of open systems for optimization of the exergy loss is an important subject in the course.

5 Thermodynamics in Living Systems

Due to the irreversible nature of life processes, they invariably and continuously dissipate Gibbs energy. As this is almost always reflected in a continuous release of heat, the phenomenon can be monitored in a calorimeter. The possible implications and applications of this dual dissipation of heat and Gibbs energy are also presented in the course.

Heat effects in cellular cultures often go unnoticed when one is working with conventional laboratory equipment because most of the heat release by the culture is lost to the environment too quickly to give rise to a perceivable temperature increase. This, however, is completely different on a large scale [47]. As opposed to laboratory reactors, industrial size fermenters operate nearly adiabatically due to their much smaller surface to volume ratio. Thus, all the heat released by the culture must be removed by appropriate cooling facilities. It is therefore of great practical importance to have sufficient quantitative information on microbial heat release when designing the cooling facilities for biotechnological processes.

The continuous generation of heat by microbial cultures can also be used as a basis for an on-line monitoring of the microbial activity and metabolism. If the temperature increase in the cooling water, its flow rate, and the other relevant energy exchange terms such as agitation and evaporation rates are measured systematically, the heat dissipation rate of the cellular culture can quantitatively be monitored on-line in industrial fermenters. The information contained in this signal can be used to optimize the bioprocess and for on-line process control.

This has clearly been demonstrated at the laboratory [48, 49], as well as at the industrial scale [50]. Monitoring heat generation rates of microbial and animal cell cultures at the laboratory scale can yield extremely valuable additional information on the state of the culture and on metabolic events [51–53], but this potential is only rarely exploited.

The continuous heat generation that is so typical of life reflects, as already stated, the continuous need for free energy dissipation. Figure 8 shows a simple explanation of this need for a growing cell culture. The biosynthesis of biopolymers, membranes, functional structures organelles, and all the other highly complex items of which a living cell consists, from simple molecules such as carbohydrates and simple salts, is most often endergonic due to entropic reasons. To

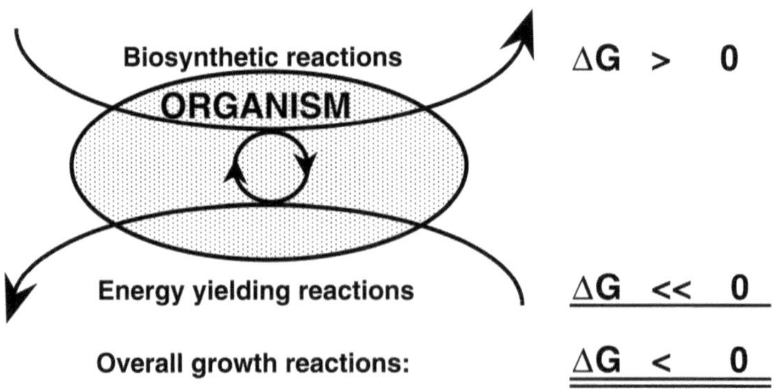

Fig. 8. Biosynthesis and Gibbs energy dissipation in cellular systems

drive all these biosynthetic reactions despite the increase of ΔG, they are coupled, in chemotrophic organisms, to one or several catabolic or "energy-yielding" reactions. The latter are highly exergonic such that the overall growth reaction occurs spontaneously to such a degree that it is essentially irreversible.

Free energy dissipation and growth yield are obviously related. If a large amount of free energy must be dissipated to drive the biosynthesis of a given amount of biomass, the growth yield will be small, but both the heat generation and the Gibbs energy dissipation per amount of biomass will be substantial. If, on the other hand, the metabolism gets away with only modest energy dissipation for the same growth, there will only be a small heat effect, but the growth yield will be large. The upper limit of the growth yield is given by an idealized equilibrium growth process, in which the free energy changes of the biosynthetic and the energy yielding reaction just cancel each other so that the overall dissipation of Gibbs energy is zero. Real growth processes are, however, far away from this limit.

A thermodynamic analysis obviously offers potential as a basis for predicting growth yields. Several correlations have been proposed comparing actual growth stoichiometries with the upper limit just described in terms of thermodynamic efficiencies [54, 55].

By far the most complete of these correlations is by Heijnen and coworkers [56]. It is based on a large body of literature and correlates the overall Gibbs energy dissipation as well as the maintenance requirements in terms of simple variables such as the number of carbon atoms and the degree of reduction of the carbon and energy source, respectively [56, 57]. From this prediction of the overall Gibbs energy dissipation, the growth yield may be calculated based on simple energy balances [57, 58].

The analysis of Gibbs energy dissipation yields insight into the thermodynamics of living systems. It may be stated that microorganisms by and large need to dissipate about 300–500 kJ of Gibbs energy per C-mol of biomass grown, but in special cases the figure may exceed 1000 kJ C-mol^{-1} [56]. Although catabolism provides the driving force for growth and therefore is responsible for Gibbs energy dissipation, microorganisms use various thermodynamic strategies for attaining the necessary amount of dissipation. The overall ΔG may be negative because of a negative ΔH or a positive $T\Delta S$:

$$\Delta G = \Delta H - T\Delta S \tag{1}$$

Depending on which term in Eq. (1) is dominating, growth is said to be enthalpy- or entropy-driven [58].

Respiration is a case of enthalpy-driven growth. The change of entropy stored in all chemicals when substrates are transformed into biomass, CO_2, and water, as reflected in ΔS, is nearly zero, and the Gibbs energy change is almost equal to ΔH. This is the reason why respiratory growth processes are fairly exothermic. In fermentative processes, however, the enthalpy change is not nearly as negative, since no external electron acceptor is involved. However, fermentative catabolic reactions degrade the energy substrate into many smaller molecules so that ΔS is highly positive despite the fact that it includes the formation of a small amount of biomass that has a low entropy. Fermentations are thus essentially entropy-driven. Some fermentations yield such a highly positive $T\Delta S$ term that ΔG is negative, and the cells grow despite the fact that ΔH is positive, which means that they are forced to produce fermentative waste products containing more energy than the energy substrate. It has been confirmed calorimetrically that such growth processes are endothermic, that is, that such cells cool their environment while growing [59].

All these analyses are based on a simple black box approach. As has been mentioned, such analyses are highly useful for predicting biomass yields and microbial stoichiometry based on a minimal amount of information. On the other hand, they cannot predict very well the yields of non-catabolic metabolites nor indicate whether and how product yields could be improved. For this, the black box must be opened and a more detailed analysis of the metabolism has to be performed. First ideas for a thermodynamic analysis of metabolic pathways have been published by some authors [60–62].

However, much research remains to be done in this area. The thermodynamics of metabolic flux analysis has not yet been well established and free energy loss analysis based on metabolic flux analysis has only been applied to some particular problems, although there might be room for the development of a systematic methodology.

6 Conclusions

The development of a rigorous thermodynamic description of the excruciatingly complex world of biotechnology may seem a daunting task but is also one of the major challenges in establishing the scientific basis for rational, efficient, and rapid bioprocess development and design. Quite a body of knowledge exists already, but a wider use of many branches such as thermodynamics of charged biopolymers, correlative approaches, and thermodynamics for open and irreversible systems, needs to be encouraged, for example, by advanced courses such as the one described here. But further research is needed into many different areas. They include increasing our base of reliable data on phase equilibria and on free energy of biomolecules in their environment, with a particular emphasis on not only proteins but also DNA and other biopolymers, further developing both theoretical and correlative approaches, research into thermodynamics effects in biopolymer stability and function, application of classical and irreversible thermodynamics to cellular systems, large-scale biocalorimetry, energy and free energy loss analysis of whole biotechnological processes, cellular growth processes, and metabolic schemes. The scope for novel research into these and many other related areas is enormous and the results are essential to meet the challenge outlined above.

Acknowledgement. Financial support of the European Science Foundation through its programme Process Integration in Biochemical Engineering is gratefully acknowledged.

7 References

1. Luyben KCAM, van der Wielen LAM (1993) ESF program on Process Integration in Biochemical Engineering. Reports on the Workshop on Integrated Downstream Processes, Delft, 11 14 February 1993
2. von Stockar U, van der Wielen LAM (1997) Thermodynamics in biochemical engineering. J Biotechnol 59:25–37
3. Prausnitz JM, Lichtenthaler RN, de Azevedo EG (1986) Molecular thermodynamics of fluid-phase equilibria. Prentice-Hall, Englewood Cliffs, NJ
4. Prausnitz JM (1995) Some new frontiers in chemical engineering thermodynamics. Fluid Phase Equilib 104:1–20
5. Chen CC, Zhu Y, Evans LB (1989) Phase partitioning of biomolecules: Solubilities of amino acids. Biotechnol Prog 5(3):111–118
6. Orella CJ, Kirwan DJ (1991) Correlation of amino acid solubilities in aqueous aliphatic alcohol solutions. Ind Eng Chem Res 30(5):1040–1045
7. Coen C, Chiew YC, Blanch HW, Prausnitz JM (1996) Salting out of aqueous proteins: Phase equilibria and intermolecular potentials. AIChE J 41(4):996–1004
8. Gude MT, van der Wielen LAM, Luyben KCAM (1996a) Phase behavior of α-amino acids in multicomponent aqueous alkanol solutions. Fluid Phase Equilib 116:110–117
9. Gude MT, Meuwisen HHJ, van der Wielen LAM, Luyben KCAM (1996) Partition coefficients and solubilities of α-amino acids in aqueous 1-butanol solutions. Ind Chem Eng Res 35:4700–4712
10. Khoskbarchi M, Vera JH (1996) Measurements of activity coefficients of amino acids in aqueous electrolyte solutions: experimental data for the system H_2O+NaCl+glycine and H_2O+NaCl+DL-alanine at 25 °C. Ind Eng Chem Res 35:2735–2742
11. George A, Wilson WW (1994) Predicting protein crystallization from a dilute-solution property. Acta Crystallogr D50:361–365
12. Eiteman MA, Gainer JL (1991) Predicting partition coefficients in polyethylene glycol-potassium phosphate aqueous two-phase systems. J Chrom 27(3):313–324
13. Thien MP, Hatton TA, Wang DIC (1988) Separation and concentration of amino acids using liquid emulsion membranes. Biotechnol Bioeng 32:604–615
14. el Rassi Z, Lee AN, Horvath C (1990) Reversed-phase and hydrophobic interaction chromatography of peptides and proteins. In: Asenjo JA (ed) Separation processes in biotechnology. Dekker, NY, pp 447–494
15. Jones IL, Carta G (1993) Ion exchange of amino acids and dipeptides on cation resins with various degree of crosslinking. 1. Equilibrium. Ind Eng Chem Res 32:107–117
16. Bowen WR, Moran EB (1995) Ion exchange of amino acids at a natural organic ion exchanger: Thermodynamics and energetics. Biotechnol Bioeng 48:559–572
17. van Berlo M, Gude MT, van der Wielen LAM, Luyben KCAM (1997) Solubilities and partitioning of glycine in water+ethanol+butanol solutions. Ind Eng Chem Res 36(6): 2474–2482
18. Albertsson P-A, Johansson G, Tjerneld F (1990) Aqueous two-phase separations. In: Asenjo JA (ed) Separation processes in biotechnology. Dekker, NY, pp 287–328
19. Glatz CE (1990) Precipitation. In: Asegno JA (ed) Separation processes in biotechnology. Dekker, Basel, pp 329–356
20. Kim DW, Jeong YK, Lee JK (1994) Adsorption kinetics of exoglucanase in combination with endoglucanase from *T. viride* on microcrystalline cellulose and its influence on synergistic degradation enzyme. Microb Techn 16:649–658
21. Schmidt AS, Andrews BA, Asenjo JA (1996) Correlation for the partition behaviour of proteins in ATPS: Effect of overall protein concentration. Biotechnol Bioeng 50: 617–626
22. Evans DF, Wennerström H (1994) The colloidal domain. Where physics, chemistry, biology and technology meet. VCH, NY

23. Jansen ML, Straathof AJJ, van der Wielen LAM, Luyben KCAM, van den Tweel WJJ (1996) A rigorous model for ion exchange equilibria of strong and weak electrolytes AIChE J 42(7):1911–1924
24. Jansen ML, Hofland GW, Houwers J, Straathof AJJ, van der Wielen LAM, Luyben KCAM, van den Tweel WJJ (1996) Effect of pH and concentration on column dynamics of weak electrolyte ion exchange processes. AIChE J 42(7):1925–1937
25. van der Wielen LAM, Lankveld MJA, Luyben KCAM (1995) Anion exchange equilibria of penicillin G, phenylacetic acid and 6-aminopenicillanic acid versus Cl^- in IRA400 ion exchange resin. J Chem Eng Data 41:239–243
26. van der Wielen LAM, Jansen ML, Luyben KCAM (1996) Multicomponent ion exchange equilibria of weak electrolyte biomolecules. In: Greig JA (ed) Ion exchange: developments and applications. SCI, Cambridge, UK, pp 292–290
27. Bisschops MAT, van der Wielen LAM, Luyben KCAM (1998) Swelling kinetics of gel-type particles using Maxwell Stefan theory. Ind Eng Chem Res 37(8):3312–3322
28. Montbriand PM, Malone RW (1996) Improved method for the removal of endotoxin from DNA. Biotechn J 44:43–46
29. Blasey HD, Aubry J-P, Mazzei GJ, Bernard AR (1996) Large scale transient expression with COS cells. Cytotechnology 18:183–192
30. Jordan M, Schallhorn A, Wurm FM (1996) Transfecting mammalian cells: optimization of critical parameters affecting calcium-phosphate precipitate formation. Nucleic Acids Res 24:596–601
31. Ulmer JB, Donnelly JJ, Parker SE, Rhodes GH, Felgner PL, Dwarki VJ, Gromkowski SH, Deck RR, DeWitt CM, Friedman A, Hawe LA, Leander KR, Martinez D, Perry HC, Shiver JW, Montgomery DL, Liu MA (1993) Heterologous protection against influenza by injection of DNA encoding a viral protein. Science 259:1745–1749
32. Tramper J, Vermüe MH, Beeftink HH, von Stockar U (eds) (1992) Biocatalysis in non-conventional media. Progress in biotechnology 8, Proceedings of an International Symposium. Noordwijkerhout, 26–29 April 1992, Elsevier, Amsterdam
33. Halling PJ (1994) Thermodynamic predictions for biocatalysis in non-conventional media: theory, tests, and recommendations for experimental design and analysis. Enzyme Microb Technol 16:178–206
34. von Stockar U (1992) Concluding remarks. In: Tramper J et al. (eds) Biocatalysis in non-conventional media. Progress in biotechnology 8:201–206, Elsevier, Amsterdam
35. Erbeldinger M, Ni X, Halling J (1998) Enzymatic synthesis with mainly undissolved substrates at very high concentrations. Enzyme Microb Technol 23:141–148
36. Halling PJ, Eichhorn U, Kuhl P, Jakubke HD (1995) Thermodynamics of solid-to-solid conversion and application to enzymic peptide synthesis. Enzyme Microb Technol 17: 601–606
37. Janssen A (1993) Enzymatic synthesis of polyol esters in aqueous-organic two-phase systems. PhD thesis, Agricultural University of Wageningen, NL
38. Wilke CR, von Stockar U (1991) Absorption. In: Kirk-Othmer encyclopedia of chemical technology 1:38–94, 5th edn, Wiley, NY
39. Maxwell JC (1867) Phil Trans Roy Soc 157:49–88. Reprinted in Scientific Papers (1962) Dover, NY, 1:392–409
40. Stefan J (1871) Sitzber Akad Wiss Wien 63:63–124
41. Wesselingh JA, Krishna R (2000) Mass transfer in multicomponent mixtures. Delft University Press, Delft, The Netherlands
42. Wei IC, Rowley RL (1985) A local composition model for multicomponent liquid mixture shear viscosity. Chem Eng Sci 40:401–408
43. Vignes A (1966) Diffusion in binary liquids. Ind Eng Chem Fundam 5:189–199
44. Le Goff P (1979–1981) Energétique Industriélle. Technique et Documentation Lavoisier, Paris, 3 vol
45. de Swaan Arons J, van der Kooij HJ (1993) Exergy analysis. Adding insight and percision to experience and intuition. In: Weijnen MPC, Drinkenburg B (eds) Precision process technology, Kluwer, Dordrecht, pp 89–114

46. Sheldon R (1993) The role of catalysis in waste minimization. In: Weijnen MPC, Drinkenburg AAH (eds) Precision process technol, Kluwer, pp 125–138
47. von Stockar U, Marison IW (1991) Large-scale calorimetry and biotechnology. Thermochim Acta 193:215–242
48. Randolph TW, Marison IW, Martens DE, von Stockar U (1990) Calorimetric control of fed-batch fermentations. Biotechnol Bioeng 36:678–684
49. Larsson C, Lidén G, Niklasson C, Gustafsson L (1991) Calorimetric control of fed batch cultures of *Saccharomyces cerevisiae*. Bioprocess Eng 7:151–155
50. Voisard D, Pugeaud P, Kumar AR, Jenny K, Jayaraman K, Marison IW, von Stockar U (2002) Development of a large scale biocalorimeter to monitor and control bioprocesses. Biotechnol Bioeng (in press)
51. Randolph TW, Marison IW, Berney C, von Stockar U (1989) Bench scale calorimetry of hybridomas in suspension culture. Biotech Technol 3:369–374
52. Duboc P, Marison IW, von Stockar U (1996) Physiology of *Saccharomyces cerevisiae* during Cell Cycle Oscillations. J Biotechnol 51:57–72
53. Garcia Payo MC, Ampuero S, Oviedo A, Liu J-S, van Gulik W, Marison IW, von Stockar U (2002) Development and characterization of a bio-reaction calorimeter for animal cell cultures. Thermochim Acta (in press)
54. Roels JA (1983) Energetics in biotechnology, Elsevier, Amsterdam
55. Westerhoff HV, van Dam K (1987) Thermodynamics and control of biological free-energy transduction. Elsevier, Amsterdam
56. Heijnen JJ, van Loosdrecht MCM, Tijhuis L (1992) A black-box mathematical model to calculate auto- and heterotrophic biomass yield based on Gibbs energy dissipation. Biotechnol Bioeng 40:1139–1154
57. Tijhuis L, van Loosdrecht M, Heijnen JJ (1993) A thermodynamically based correlation for maintenance Gibbs energy requirements in aerobic and anaerobic chemotrophic growth. Biotechnol Bioeng 42:509–519
58. Liu JS, von Stockar U (1999) Does microbial life always feed on negative entropy? Thermodynamic analysis of microbial growth. Biochem Biophys Acta 1412:191–211
59. Liu JS, Marison IW, von Stockar U (2001) Microbial growth by a net heat up-take: a calorimetric and thermodynamic study on acetotrophic methanogenesis by *Methanosarcina barkeri*. Biotechnol Bioeng 75:170–180
60. Mavrovouniotis ML (1993) Identification of localized and distributed bottlenecks in metabolic pathways. ISMB-93:273–283
61. Pissarra PD, Nielsen J (1997) Thermodynamics of metabolic pathways for penicillin production: analysis of thermodynamic feasibility and free-energy changes during fed-batch cultivation. Biotechnology Prog 13:156–165
62. Stephanopoulos GN, Aristidou AA, Nielsen J (1998) Metabolic engineering. Principles and methodologies. Academic Press, San Diego

Received: March 2002

Integration of Physiology and Fluid Dynamics

Sven Schmalzriedt · Marc Jenne · Klaus Mauch · Matthias Reuss

Institute of Biochemical Engineering, University of Stuttgart, Allmandring 31,
70569 Stuttgart, Germany. *E-mail: reuss@ibvt.uni-stuttgart.de*

The purpose of strategies for the integration of fluid dynamics and physiology is the development of more reliable simulation tools to accelerate the process of scale-up. The rigorous mathematical modeling of the richly interactive relationship between the dynamic response of biosystems and the physical environment changing in time and space must rest on the link between coupled momentum, energy and mass balances and structured modeling of the biophase. With the exponential increase in massive computer capabilities hard- and software tools became available for simulation strategies based on such holistic integration approaches. The review discusses fundamental aspects of application of computational fluid dynamics (CFD) to three-dimensional, two-phase turbulence flow in stirred tank bioreactors. Examples of coupling momentum and material balance equations with simple unstructured kinetic models for the behavior of the biophase are used to illustrate the application of these strategies to the selection of suitable impeller configurations. The examples reviewed in this paper include distribution of carbon and energy source in fed batch cultures as well as dissolved oxygen fields during aerobic fermentations.

A more precise forecasting of the impact of the multitude of interactions must, however, rest upon a rigorous understanding of the response of the cell factory to the complex dynamic stimulation due to space- and time-dependent concentration fields. The paper also introduces some ideas for fast and very fast experimental observations of intracellular pool concentrations based on stimulus response methods. These observations finally lead to a more complex integration approach based on the coupling of CFD and structured metabolic models.

Keywords. Computational fluid dynamics (CFD), Intracellular metabolites, Integration of CFD with unstructured and structured kinetic models

Advances in Biochemical Engineering/
Biotechnology, Vol. 80
Series Editor: T. Scheper

Abbreviations

a	m^{-1}	specific interfacial surface area
$a_{vm,i}$	$m\ s^{-2}$	array of virtual acceleration
A_b	m^2	sectional area of a bubble
c	–	parameters of turbulence models
c_b	–	constant in calculation of bubble diameter
c_d	–	drag coefficient
c_k	$mol\ m^{-3}$ or $g\ m^{-3}$	concentration of species k
$c^*_{O_2}$	$mol\ m^{-3}$	oxygen concentration at the gas-liquid interface
c_{vm}, c_{vma}	–	coefficients of virtual mass force
c_μ	–	parameter of turbulence model
$c_{\mu,b}$	–	parameter for calculation of bubble induced turbulence
d_b	m	bubble diameter
d_i	m	impeller diameter
D	h^{-1}	dilution rate
D_{eff}	$m^2\ s^{-1}$	turbulent dispersion coefficient
D_{O_2}	$m^2\ s^{-1}$	diffusion coefficient of oxygen
D_T	m	tank diameter
f_k	–	correction factor of drag coefficient
F	N	force
F_d	N	drag force
g	$m\ s^{-2}$	gravitational acceleration
H	m	liquid height
H	bar	Henry-number
I	–	inhomogeneity
k	$m^2\ s^{-2}$	turbulent kinetic energy
k_L	$m\ s^{-1}$	mass transfer coefficient
k_La	s^{-1}	volumetric mass transfer coefficient
K_k	$mol\ m^{-3}$ or $g\ m^{-3}$	half saturation constant of species k
n	s^{-1}	impeller speed
n_t	–	number of tanks

p	bar	pressure
p_{O_2}	bar	partial pressure of oxygen in the gas phase
P	W	power input of the impeller
P_k	$m^2\ s^{-3}$	production of turbulent kinetic energy
Q_L	$m^3\ s^{-1}$	liquid pumping capacity of an impeller
q_k	s^{-1}	specific rate of species k
r	m	radial coordinate
r_k	$mol\ m^{-3}\ s^{-1}$ or $g\ m^{-3}\ s^{-1}$	reaction rate of species k
S_k	$mol\ m^{-3}\ s^{-1}$ or $g\ m^{-3}\ s^{-1}$	source of species k
S_i	$N\ m^{-3}$	specific force
Sc	–	Schmidt number
Sc_t	–	turbulent Schmidt number
t	s	time
t_c	s	circulation time
u_i	$m\ s^{-1}$	mean velocity component
u_i'	$m\ s^{-1}$	fluctuation velocity component
V	m^3	volume
$\dot{V}_G$	$m^3\ s^{-1}$	gas sparging rate
V_T	m^3	tank volume
x	m	coordinate
x_{O_2}	–	concentration fraction of oxygen in the liquid phase
y_{O_2}	–	molar fraction of oxygen in the gas phase
Y	–	yield coefficient
z	m	axial coordinate

Greek letters

δ	–	Kronecker symbol
ε	$m^2\ s^{-3}$	energy dissipation rate
ε_G	–	volume fraction of gas phase
ε_L	–	volume fraction of liquid phase
φ	°	tangential coordinate
μ	h^{-1}	specific growth rate
μ	Pa s	dynamic viscosity
μ_m	Pa s	modified dynamic viscosity
$\nu_{eff} = \nu_L + \nu_t$	$m^2\ s^{-1}$	effective viscosity
ν_L	$m^2\ s^{-1}$	laminar viscosity
ν_t	$m^2\ s^{-1}$	turbulent viscosity
t_m	s	mixing time
ϱ	$kg\ m^{-3}$	density
σ	$N\ m^{-1}$	surface tension
σ	–	parameters of turbulence models
τ_d	s	dissipation-range timescale
τ_{ij}	–	laminar deformation tensor
τ_p	s	production-range timescale

Subscripts

0	without gassing
Ac	Acetoin
Bu	Butanediol
d	drag
G	gas phase
L	liquid phase
O_2	oxygen
P	product
S	substrate
vm	virtual mass
X	biomass

Dimensionless numbers

$Mo = (\varrho_L - \varrho_G)\, g\, \mu_L^4\, \varrho_L^{-2}\, \sigma^{-3}$	Morton number
$Re = \lvert \Delta u \rvert\, d_b\, v_L^{-1}$	Reynolds number of a bubble
$Sc = v_L\, D_{O_2}^{-1}$	Schmidt number
$Sc_t = v_{\text{eff}}\, D_{\text{eff}}^{-1}$	turbulent Schmidt number
$We = \varrho_L\, \lvert \Delta u \rvert^2\, d_b\, \sigma^{-1}$	Weber number

1 Introduction

The physiological state of cellular systems and its related behavior with respect to growth and product formation is the result of a complex interplay between the extracellular environment and the cellular machinery. Functionality of a biosystem for the purpose of bioproduction processes is therefore determined by the co-operative actions of the extracellular stimuli and functional genomics (Fig. 1).

Engineering of optimal reactors in which living cells function as the factory is further complicated because of the dynamic variations of the extracellular environment. A quantitative description of these phenomena should consequently rest upon the two interwoven aspects of structured bioprocess modeling (Fig. 2). The first aspect concerns the complex interaction of the functional units of the cells, including the mathematical formulation of reaction rates and the key regulation of these networks in response to changes in the environment. The second aspect has to do with the structure of the abiotic phases of the bioreactor in order to analyse the quality of mixing and other transport phenomena between the phases causing gradients in the concentrations of various substrates and products.

These problems are particularly important for those processes in which nutrients are continuously introduced into the broth. For specific nutrients such as oxygen and sometimes other nutrients such as carbon source, the time constant for their distribution (mixing-time) may be of the same magnitude as those of their consumption in any reasonable sized reactor beyond bench-scale. If we accept that spatial variations exist, we are faced with the problem of dynamically

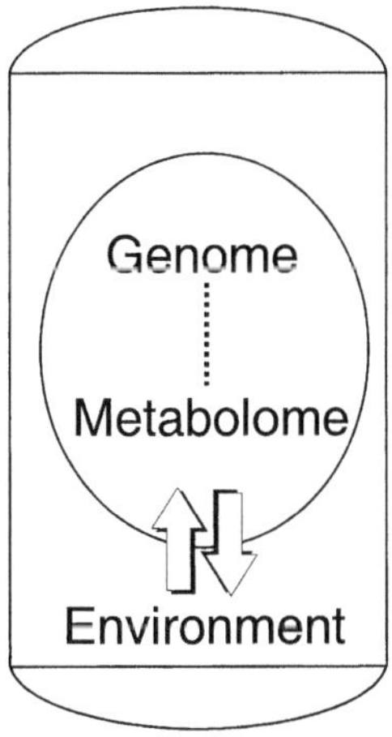

Fig. 1. Extracellular stimuli and functional genomics

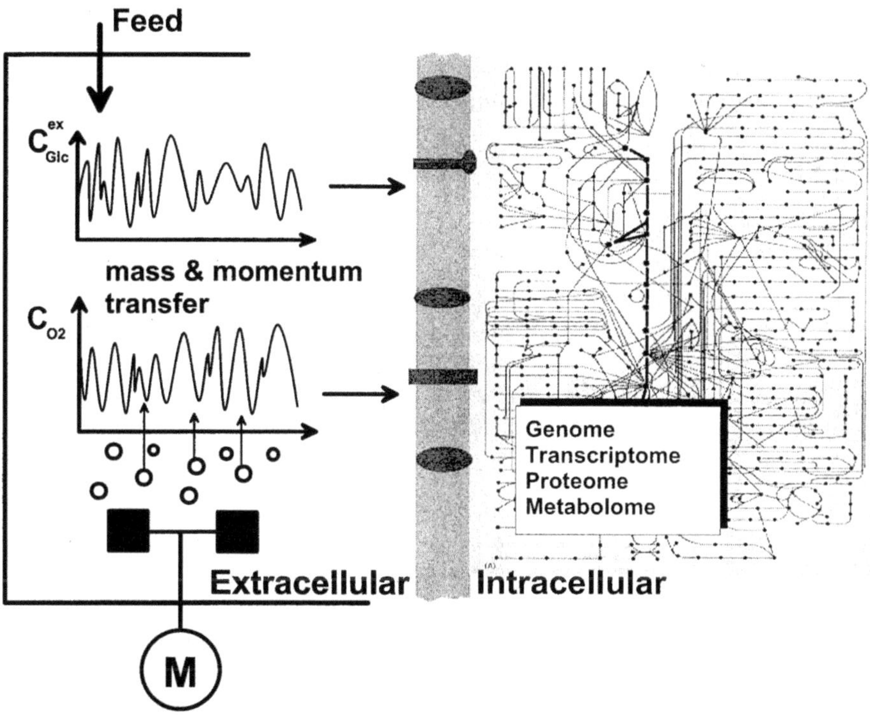

Fig. 2. Aspects of bioprocess modeling

changing environmental conditions. This in turn may result in drastic changes in metabolism and final outcome of the process. The long-term mathematical description of these phenomena requires flexible tools to be adapted to different systems and to be able to integrate the process and reactor.

Successful strategies to be developed for this challenging task require bridging disciplines of engineering and molecular biochemistry. Accordingly, this paper addresses these issues by discussing the following tools:

(1) Application of computational fluid dynamics (CFD) for modeling and simulation of the flow behavior of the abiotic phases.
(2) Coupling of material balance equations for carbon and energy source as well as oxygen with fluid dynamics considering unstructured rate expressions.
(3) Experimental observations and structured modeling of fast intracellular response to dynamic disturbances.
(4) Coupling of intracellular reaction with extracellular concentration fields.

2 Modeling and Simulation of Gas-Liquid Flow in Stirred Tank Reactors

It is generally now accepted that Reynolds-averaging Navier-Stokes equations and modeling the Reynolds-stresses with an appropriate turbulence model is a promising method of flow behavior modeling. Ongoing development of commercial computational fluid dynamics software (CFD) and increasing computer power are continuously improving the conditions for the simulation of the three-dimensional and turbulent flow structure in stirred tanks.

2.1 Liquid Flow

Among the variety of impellers, the Rushton turbine is well established for many tasks, mainly due to good gas dispersion and mixing of liquids with low viscosities. The Rushton turbine generates a flow leaving the impeller in radial and tangential directions. This radial-tangential jet flow divides at the vessel wall and the flow then recirculates back into the impeller region. Besides turbulent dispersion, recirculation of the flow is the main reason for the mixing capability of stirred tanks.

In spite of improved hard- and software, which have greatly expanded the tools available for simulating fluid flow in stirred tank reactors, a number of unsolved problems and open questions still exist.

A critical analysis of the many publications concerning the simulation of liquid flow in baffled stirred tank reactors equipped with a Rushton turbine reveals several discrepancies. The most important differences between the simulations concern the dimensionality of the simulations (three-dimensional or axisymmetric), turbulence modeling, the modeling approaches for the Rushton turbine as well as the accuracy of the numerical predictions, which depends on the grid size.

The different modeling approaches for the single-phase flow with a Rushton turbine have been examined and critically reviewed by Jenne and Reuss [1]. In what follows, only the basic principles will be summarized.

The transport equations describing the instantaneous behavior of turbulent liquid flow are three Navier-Stokes equations (transport of momentum corresponding to the three spatial coordinates r, z, φ in a cylindrical polar coordinate system) and a continuity equation. The instantaneous velocity components and the pressure can be replaced by the sum of a time-averaged mean component and a root-mean-square fluctuation component according to Reynolds. The resulting Reynolds equations and the continuity equation are summarized below:

$$\frac{\partial(\varrho u_i)}{\partial t}+\frac{\partial(\varrho u_i)}{\partial x_i}=-\frac{\partial}{\partial x_i}\left(\tau_{ij}+\varrho u_i'\right)-\frac{\partial p}{\partial x_i}+\varrho g_i \tag{1}$$

$$\frac{\partial \varrho}{\partial t}+\frac{\partial}{\partial x_i}\left(\varrho u_i\right)=0 \tag{2}$$

A reasonable compromise for model accuracy and computational expense are eddy viscosity models relating the individual Reynolds stresses to mean flow gradients:

$$\varrho u_i' u_j'=-\varrho \upsilon_i\left(\frac{\partial u_i}{\partial x_j}+\frac{\partial u_j}{\partial x_i}\right)+\frac{2}{3}\varrho \delta_{ij} k \tag{3}$$

Where υ_t is the turbulent eddy viscosity. The transport of momentum, which is related to turbulence, is thought of as turbulent eddies, which, like molecules, collide and exchange momentum.

The family of two-equation $k-\varepsilon$ models is the most widely used of the eddy viscosity models. A $k-\varepsilon$ model consists of two transport equations, one for the turbulent kinetic energy k and one for the energy dissipation rate ε. The turbulent eddy viscosity is calculated from:

$$\upsilon_t = c_\mu \frac{k^2}{\varepsilon} \tag{4}$$

where c_μ is a parameter which depends on the specific $k-\varepsilon$ model.

The standard $k-\varepsilon$ model, as presented by Launder and Spalding [2], is by far the most widely-used two-equation eddy viscosity model, also for modeling turbulence in stirred tank reactors. The popularity of the model and its wide use and testing has thrown light on both its capabilities and its shortcomings, which are well documented in the literature [2–8]. For high turbulent Reynolds numbers, the model may be summarized as follows:

$$\frac{\partial(\varrho k)}{\partial t}+\frac{\partial}{\partial x_i}\left(\varrho u_i k\right)=\frac{\partial}{\partial x_i}\left(\varrho \frac{\upsilon_{\text{eff}}}{\sigma_{k,S}}\frac{\partial k}{\partial x_i}\right)+\varrho\left(P_k-\varepsilon\right) \tag{5}$$

$$\frac{\partial(\varrho \varepsilon)}{\partial t}+\frac{\partial}{\partial x_i}\left(\varrho u_i \varepsilon\right)=\frac{\partial}{\partial x_i}\left(\varrho \frac{\upsilon_{\text{eff}}}{\sigma_{\varepsilon,S}}\frac{\partial \varepsilon}{\partial x_i}\right)+\varrho\left(c_{1,S}\frac{\varepsilon}{k}P_k-c_{2,S}\frac{\varepsilon}{k}\varepsilon\right) \tag{6}$$

The model parameters of the standard $k-\varepsilon$ model are listed in Table 1.

Table 1. Parameters of the standard $k-\varepsilon$ model

Parameter	$c_{\mu,S}$	$c_{1,S}$	$c_{2,S}$	$\sigma_{k,S}$	$\sigma_{\varepsilon,S}$
Value	0.09	1.44	1.92	1.00	1.314

The production of turbulent kinetic energy P_k is modeled with the aid of the eddy viscosity hypothesis:

$$P_k = \nu_t \left(\frac{\partial u_i}{\partial x_j} + \frac{\partial u_j}{\partial x_j} \right) \frac{\partial u_i}{\partial x_j} \tag{7}$$

The dissipation rate ε can be regarded as the rate at which energy is being transferred across the energy spectrum from large to small eddies. The standard $k-\varepsilon$ model assumes spectral equilibrium, which implies that once turbulent kinetic energy is generated at the low-wavenumber end of the spectrum (large eddies), it is dissipated immediately at the same location at the high-wavenumber end (small eddies). In other words, the standard $k-\varepsilon$ model assumes that P_k is near to ε. As far as the stirred vessel is concerned, this is a very restrictive assumption, because there is a vast size disparity between those eddies in which turbulence production takes place (mainly at the stirrer), and the eddies in which turbulence dissipation occurs.

The standard $k-\varepsilon$ model employs a single time scale $\tau_d = k/\varepsilon$, called dissipation range timescale, in the ε equation to characterize the dynamic processes occurring in the energy spectrum. Thus, Eq. (6) can be rewritten as:

$$\frac{\partial (\rho \varepsilon)}{\partial t} + \frac{\partial}{\partial x_i} (\rho\, u_i\, \varepsilon) = \frac{\partial}{\partial x_i} \left(\rho \frac{\upsilon_{\text{eff}}}{\sigma_{\varepsilon,S}} \frac{\partial \varepsilon}{\partial x_i} \right) + \rho \left(c_{1,S} \frac{P_k}{\tau_d} - c_{2,S} \frac{\varepsilon}{\tau_d} \right) \tag{8}$$

The energy spectrum, however, comprises fluctuating motions with a spectrum of timescales, and a single timescale approach is unlikely to be adequate under all circumstances. Consequently, the model has been found to perform less satisfactorily in a number of flow situations, including separated flows, streamline curvature, swirl, rotation, compressibility, axisymmetrical jets, etc.

Because the model is so widely used, variants and ad hoc modifications aimed at improving its performance abound in the literature. The most well-known modifications are the Chen-Kim and RNG variant of the $k-\varepsilon$ model.

To ameliorate the previously mentioned deficiencies in the standard $k-\varepsilon$ model, Chen and Kim [9] proposed a modification, which improves the dynamic response of the ε equation by introducing an additional timescale

$$\tau_P = \frac{k}{P_k} \tag{9}$$

which is called the production-range timescale. The final expression of the transport equation for the dissipation rate is given as:

$$\frac{\partial(\varrho\varepsilon)}{\partial t}+\frac{\partial}{\partial x_i}(\varrho u_i\varepsilon)=\frac{\partial}{\partial x_i}\left(\varrho\frac{\nu_{\text{eff}}}{\sigma_{\varepsilon,CK}}\frac{\partial\varepsilon}{\partial x_i}\right)+\varrho\left(\underbrace{\overbrace{c_{1,CK}\frac{P_k}{\tau_d}}^{\text{1st part}}+\overbrace{c_{3,CK}\frac{P_k}{\tau_p}}^{\text{2nd part}}}_{\text{production term}}-c_{2,CK}\frac{\varepsilon}{\tau_d}\right) \tag{10}$$

The parameters of the Chen-Kim model are summarized in Table 2.

The first part of the production term corresponds with the production term of the standard $k-\varepsilon$ model. Notice that the second production term is related to the time scale τ_p. The introduction of this additional term enables the energy transfer to respond more efficiently to the mean strain than the standard $k-\varepsilon$ model does. Thus, τ_p enables the development of a field of ε suppressing the well-known overshoot phenomenon of the turbulent kinetic energy k. This overshoot appears, when the standard $k-\varepsilon$ model is applied to flow conditions with large values of mean strain [4, 7, 8].

The modification may be summarized as follows: ε production appears in two energy fluxes divided by two different timescales τ_d and τ_p. The multiplying coefficients might be seen as weighting factors for these two energy fluxes. One may expect that this feature offers advantages in separated flows and also in other flows in which turbulence is far from local equilibrium (P_k is far from ε). If P_k is near to ε (local equilibrium), the Chen-Kim-modified $k-\varepsilon$ model is almost identical to the standard $k-\varepsilon$ model. Then, τ_p equals τ_d, and summing up the two ε production terms leads to the ε production term of the standard $k-\varepsilon$ model. The resulting coefficient $c_{1,CK}+c_{3,CK}=1.4$ is only slightly lower than $c_{1,S}$. This is the reason why for simple boundary type flows, the Chen-Kim-modified $k-\varepsilon$ model gives results similar to those predicted by the standard $k-\varepsilon$ model. However, for complex elliptic turbulent flow problems (internal turbulent recirculating flows) involving rapid changes of turbulent kinetic energy production and dissipation rates, the Chen-Kim-modified $k-\varepsilon$ model has been shown to give much better results than the standard $k-\varepsilon$ model [9].

To further improve the agreement between simulations and experimental observations, the parameters $c_{1,CK}$ and $c_{3,CK}$ in the Chen-Kim model were slightly modified. For modification of these parameters the ratio of the Eulerian macro length scale to the impeller blade height has been employed. The property can be compared in geometric similar vessels. For more details the reader is referred to the original paper by Jenne and Reuss [1].

Table 2. Parameters of the Chen-Kim $k-\varepsilon$ model

Parameter	$c_{\mu,CK}$	$c_{1,CK}$	$c_{2,CK}$	$c_{3,CK}$	$\sigma_{k,CK}$	$\sigma_{\varepsilon,CK}$
Value	0.09	1.15	1.90	0.25	0.75	1.15

The Reynolds equations, the continuity equation, which is turned into an equation for pressure correction [10], and the transport equations for the turbulence quantities k and ε, are integrated over the respective finite volume elements resulting from the discretization of the stirred tank domain. The convection and diffusion terms in the transport equations are approximated using the hybrid-scheme of Patankar [10]. The resulting algebraic equations are then solved with the aid of the commercial CFD software PHOENICS (Version 2.1).

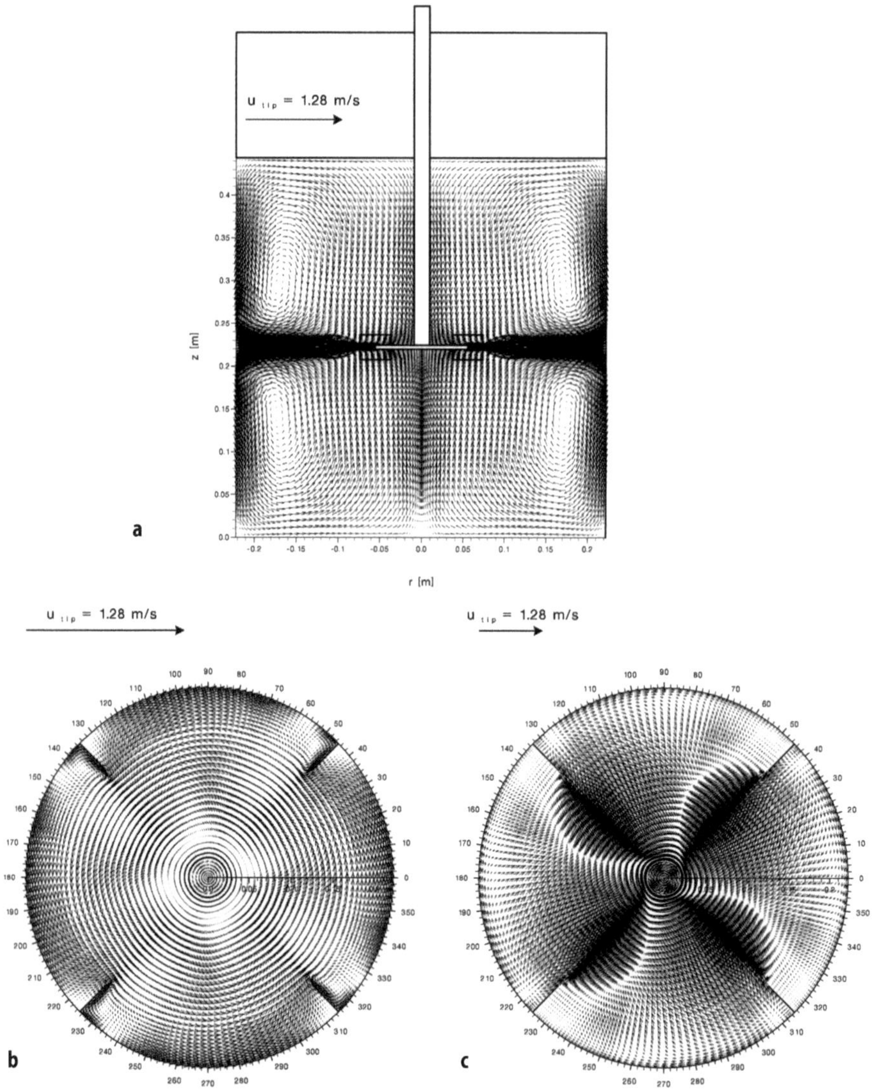

Fig. 3a–c. Simulated flow fields at a stirrer speed of 165 min^{-1}: **a** between two baffles, **b** at $z/H = 0.307$, and **c** at $z/H = 0.014$

So-called false-time-step relaxation is used to achieve stationarity. The semi-implicit method, which considers the pressure-link of the pressure correction equation and the Reynolds equations, is the SIMPLEST algorithm. The sets of algebraic equations for each variable are solved iteratively by means of the ADI technique. An example of the simulated flow field is illustrated in Fig. 3. Good agreement can then be achieved between measured flow details and the simulation results for vessels and impellers of different geometry [1].

The simulations presented here are based on experimental data for specifying the boundary conditions in the impeller, which can essentially be considered as a circumferentially and time-averaged radial-tangential jet. A resulting additional advantage is the reduced computational expense of stationary simulations compared to transient simulations. The resolution of the vortex system behind the stirrer blades (see e.g., van't Riet and Smith [11]) in applying this method, however, is not possible. To specify boundary conditions for other types of impellers one has to perform time consuming experiments in advance. To remove the two last mentioned disadvantages, recent attempts have been made to simulate the unsteady flow within and outside the impeller swept region in applying the so-called sliding-mesh technique (see e.g. Perng and Murthy [12], Takeda et al. [13]). A critical comparison of the results from the sliding mesh technique and simulations with measured data in the impeller region has been presented by Brucato et al. [14]. However, the sliding mesh technique requires excessive computational resources and for most engineering applications knowledge of the full time varying and periodic flow field may not be necessary. Another possibility to simulate flow details between the impeller blades is the so-called snapshot approach (see e.g. Ranade and Van den Akker [15]). This is often also called a multiple reference frame method [16]. Experimental data to specify boundary conditions are not necessary. An advantage compared with the sliding mesh technique is that the full time-dependent transport equations need not be solved. This offers an interesting and promising approach. However, the essential comparisons with experimental observations are lacking.

2.2 Gas-Liquid Flow

An important feature in modeling the two-phase flow is to distinguish between Eulerian and Lagrangian approaches. In the Lagrangian approach, the continuous phase is treated as a continuum while the dispersed gas bubbles are modeled as single particles. In the Eulerian approach the dispersed phase is also considered as a continuum resulting in the so-called two fluid model. Only the Eulerian approach has been considered for aerated stirred tank reactors so far. If only gravitation, pressure, and drag forces are taken into account in the momentum equation for the gas phase, the relative velocities of the gas phase are calculated from algebraic equations. This is the so-called algebraic slip model. The disadvantage of this simple approach is the fact that additional interface forces are neglected. Issa and Gosman [17] calculated the flow in a gassed and stirred vessel equipped with a Rushton turbine by using the algebraic slip model. Furthermore, they used very coarse grids because of limited computing power. Experimental verification of their simulations was not shown. Trägardh [18] reported two-dimensional simulations with the algebraic slip model for a stirred vessel equipped with three impellers. Politis et al. [19] performed three-dimensional simulations with the two fluid and $k-\varepsilon$ model. They considered different interfacial forces and critically examined their influence. These authors were able to show that in addition to the drag force, particularly the virtual mass force needs to be considered. For boundary conditions in the impeller region, values for averaged tangential velocities as well as k and ε from measured data were used.

Morud and Hjertager [20] followed an axisymmetrical approach on the two fluid and $k-\varepsilon$ model. The virtual mass force was neglected. These authors observed a considerable deviation between measured and simulated data.

The simulations of the gas-liquid flow are based on the Eulerian two fluid model originally derived by Ishii [21]. In this approach, each phase is treated as a continuum. After averaging the general transport equations, we get the following set of multi-phase conservation equations [19, 22]:

Continuity:

$$\frac{\partial}{\partial t}(\rho_k \varepsilon_k) + \frac{\partial}{\partial x_i}\left(\rho_k \varepsilon_k u_{k,i} - \rho_k \frac{\upsilon_t}{Sc_t}\frac{\partial \varepsilon_k}{\partial x_i}\right) = 0 \qquad k = L, G \tag{11}$$

A dispersive transport of gas bubbles and liquid has been considered in both continuity equations. S_{ct} is the Schmidt number for turbulent transport, which is assumed to be one [22]. The global mass conservation is given by:

$$\varepsilon_G + \varepsilon_L = 1 \tag{12}$$

MOMENTUM:
Liquid phase:

$$\frac{\partial(\rho_L \varepsilon_L u_{L,i})}{\partial t} + \frac{\partial(\rho_L \varepsilon_L u_{L,i} u_{L,j})}{\partial x_j} = -\varepsilon_L \frac{\partial}{\partial x_j}\left(\tau_{L,ij} + \rho_L u'_{L,i}\right) - \varepsilon_L \frac{\partial p}{\partial x_i} + \rho_L \varepsilon_L g_i + S_i \tag{13}$$

with the laminar shear stress $\tau_{L,ij}$ and turbulent Reynolds-stresses given by the Boussinesq approximation:

$$-\rho_L u'_{L,i} u'_{L,j} = \rho_L \nu_t \left(\frac{\partial u_{L,j}}{\partial x_j} + \frac{\partial u_{L,j}}{\partial x_i}\right) - \frac{2}{3}\rho_L \delta_{ij} k \tag{14}$$

Gas phase:

$$\frac{\partial(\rho_G \varepsilon_G u_{G,i})}{\partial t} + \frac{\partial(\rho_G \varepsilon_G u_{G,i} u_{G,j})}{\partial x_j} = -\varepsilon_G \frac{\partial p}{\partial x_j} + \rho_G \varepsilon_G g_i - S_i \tag{15}$$

Reynolds-stresses in the gas phase can be neglected.

The interfacial coupling term S_i in Eqs. (13) and (15) is a linear combination of several forces. Politis et al. [19] have compared the order of magnitude of the various forces and concluded that only the drag force and the virtual mass force need to be considered.

From the definition of the drag coefficient c_d of a single bubble, the following expression for the drag force can be derived:

$$F_{d,i} = \frac{\varrho_L}{2} c_d A_b |\Delta u| \Delta u_i \tag{16}$$

A_b is the secional area of the bubble $= (\pi/4)\, d_b^2$, Δu_i is the relative velocity between the bubble and the liquid in the direction i, $|\Delta u|$ is the absolute value of the velocity vector. The momentum equation (13) is related to the total volume dV which contains gas and liquid. The volumetric force S_i is therefore

$$S_i = \frac{F_i}{dV} = \varepsilon_G \frac{F_i}{dV_G} \tag{17}$$

and with Eq. (16):

$$S_{d,i} = \varepsilon_G \frac{3}{4} c_d \frac{\varrho_L}{d_b} |\Delta u| \Delta u_i \tag{18}$$

The correlations for air bubbles rising in distilled and tap water have been proposed by Kuo and Wallis [23]. For distilled water the equations for the drag coefficient read:

$Re < 0.49$	$c_d = 24/Re$
$0.49 < Re < 33$	$c_d = 20.68/Re^{0.643}$
$33 < Re < 661$	$c_d = 72/Re$
$661 < Re < 1237$ and $We \approx 4$	$c_d = (Re^4\, Mo)/18$
$Re > 1237$ and $We < 8$	$c_d = We/8$

For tap water, Kuo and Wallis [23] proposed the following equations:

$Re < 0.49$	$c_d = 24/Re$
$0.49 < Re < 100$	$c_d = 20.68/Re^{0.643}$
$100 < Re < 717$	$c_d = 6.3/Re^{0.385}$
$Re > 717$ and $We < 8$	$c_d = We/3$

The dimensionless numbers in these equations are defined by:

Reynolds number $$\mathrm{Re} = \frac{\varrho_L |\Delta u| d_b}{\mu_L}$$

Weber number $$We = \frac{\varrho_L |\Delta u| d_b}{\sigma}$$

Morton number $Mo = \frac{(\varrho_L - \varrho_G) g \mu_L^4}{\varrho_L^2 \sigma^3}$

Making use of the correlation proposed by Ishii and Zuber [24] for correction of the drag force in a bubble swarm, the drag coefficient c_d is multiplied with the correction factor f_k which is given by:

$$f_k = \left(\frac{1 + 17.67\, f^{\frac{6}{7}}}{18.67\, f} \right)^2 \tag{19}$$

with

$$f = \sqrt{1 - \varepsilon_G}\, \frac{\mu_L}{\mu_G} \tag{20}$$

and

$$\frac{\mu_L}{\mu_m} = (1 - \varepsilon_G)^{2.5 \frac{\mu_G + 0.4 \mu_L}{\mu_G + \mu_L}} \tag{21}$$

for $d_b > 1.8$ mm. For bubble diameters smaller than 1.8 mm, the Reynolds number is calculated with μ_m from Eq. (21).

The virtual mass force represents the force required to accelerate the apparent mass of the surrounding continuous phase in the immediate vicinity of the gas bubble. Drew and Lahey [25] have proposed the following formulation:

$$S_{vm,i} = \varrho_L\, c_{vm}\, \varepsilon_G\, a_{vm,i} \tag{22}$$

$$a_{vm,i} = \frac{\partial (u_{G,i} - u_{L,i})}{\partial t} - u_{G,i} \frac{\partial u_{G,i}}{\partial x_i} \tag{23}$$

The virtual volume coefficient c_{vm} for potential flow around a sphere is 0.5. For ellipsoidal bubbles with a ratio of semiaxes 1:2, c_{vm} is 1.12. For ellipsoidal bubbles with random wobbling motions, Lopez de Bertodano [26] calculated c_{vm} to be about 2.0. In addition, c_{vm} is a function of the specific gas holdup [27–29]:

$$c_{vm} = c_{vma} (1 - \varepsilon_G) \tag{24}$$

with

$$c_{vm} = c_{vma} (1 - 2.78 \min[0.20, \varepsilon_G]) \tag{25}$$

and

$$0.5 \le c_{vma} \le 2.0$$

For applications in two-phase flow the $k-\varepsilon$ models have been modified in different ways. One possibility is to insert additional sources into the transport equations for k and ε [30–33]. An alternative is to consider an increase of the turbulent viscosity in the liquid phase caused by the bubbles. According to Sato [34] and Lopez de Bertodano et al. [26], this effect can be described by:

$$\nu_t = c_\mu \frac{k^2}{\varepsilon} + c_{\mu,b}\, d_b\, \varepsilon_G \left|\Delta u\right| \tag{26}$$

For the parameter $c_{\mu,b}$ Lopez de Bertodano [26] suggested a value of 0.6. Assuming that the optimized version of the Chen-Kim model is still valid, the transport equations for the turbulence quantities k and ε for the two-phase system are given by:

$$\frac{\partial(\varrho_L \varepsilon_L k)}{\partial t} + \frac{\partial}{\partial x_i}\left(\varrho_L \varepsilon_L u_{L,i} k\right) = \frac{\partial}{\partial x_i}\left(\varrho_L \varepsilon_L \frac{\nu_{eff}}{\sigma_{k,s}} \frac{\partial k}{\partial x_i}\right) + \frac{\partial}{\partial x_i}\left(\varrho_L k \frac{\nu_t}{Sc_t} \frac{\partial \varepsilon_L}{\partial x_i}\right) + \varrho_L \varepsilon_L \left(P_k - \varepsilon\right) \tag{27}$$

and

$$\frac{\partial(\varrho_L \varepsilon_L \varepsilon)}{\partial t} + \frac{\partial}{\partial x_i}\left(\varrho_L \varepsilon_L u_{L,i} \varepsilon\right) = \frac{\partial}{\partial x_i}\left(\varrho_L \varepsilon_L \frac{\nu_{eff}}{\sigma_{\varepsilon,CK}} \frac{\partial \varepsilon}{\partial x_i}\right) + \frac{\partial}{\partial x_i}\left(\varrho_L \varepsilon \frac{\nu_t}{Sc_t} \frac{\partial \varepsilon_L}{\partial x_i}\right) + \varrho_L \varepsilon_L \left(c_{1,CK} \frac{P_k}{\tau_d} + c_{3,CK} \frac{P_K}{\tau_p} - c_{2,CK} \frac{\varepsilon}{\tau_d}\right) \tag{28}$$

with

$$P_k = \nu_t \left(\frac{\partial u_{L,i}}{\partial x_j} + \frac{\partial u_{L,j}}{\partial x_i}\right) \frac{\partial u_{L,i}}{\partial x_j} \tag{29}$$

As already discussed in context with the single-phase simulations, boundary conditions at the impeller are predicted from measured data of the averaged velocities. The gassed linear liquid velocities differ from the ungassed velocities because impeller power consumption and pumping capacity of the impeller decrease due to gassing.

The relation between decrease in pumping capacity and power consumption is given as

$$\frac{Q_{L,G}}{Q_{L,0}} = \left(\frac{P_G}{P_0}\right)^\alpha \tag{30}$$

with α varying between 0.34 [35, 36] and 1.0 [37]. From comparison between measured and simulated fields of specific gas hold-up discussed in the following, a value of $\alpha = 0.64$ has been estimated.

The simulations have been performed for the vessel and impeller geometries used by Bombac et al. [38, 39] in their systematic investigations of the distribution of specific gas hold-up at different speeds of agitation. These measurements were performed by using conductivity sensors. For the prediction of the interfacial forces it is necessary to estimate a representative bubble diameter. If the fluid

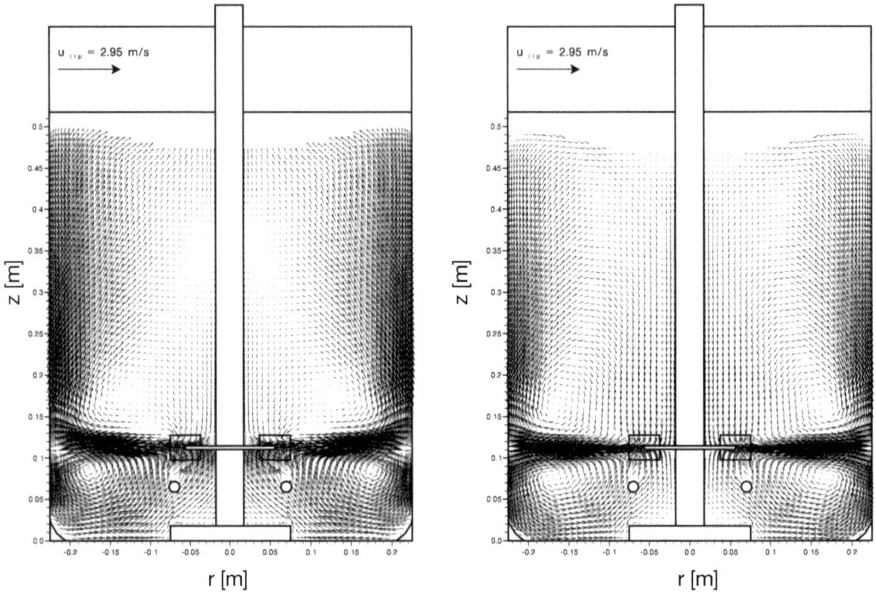

Fig. 4. Simulated flow field in the gas phase (*left*) and the liquid phase (*right*) at $n = 376\ \mathrm{min}^{-1}$ and $\dot{V}_G = 1.67 \times 10^{-3}\ \mathrm{m}^3\ \mathrm{s}^{-1}$. Tank configuration of Bombac et al. [39]

can be characterized by a hindered coalescence behavior like many fermentation broths due to their high salt concentrations, the bubble diameter is determined by the local energy dissipation in the stirrer zone. It can be calculated as the maximal stable bubble diameter according to Hinze [40]:

$$d_b = c_b \left(12 \frac{\sigma}{\varrho_L} \right)^{0.6} \varepsilon^{-0.4} \tag{31}$$

Bakker and van den Akker [41] estimated c_b to be 0.4 from bubble size data reported by Greaves and Barigou [42].

Figure 4 shows exemplarily simulated flow velocities for the gas and liquid phase. Figure 5 summarizes a comparison between simulated and predicted local values of the specific gas hold-up. For more detailed information and comparisons at different operation conditions the reader is referred to the original papers [43, 44].

2.3 Multiple Impellers

In industry, reactors are usually equipped with two or more impellers. Very little data are found concerning the details of flow patterns, particularly quantitative information about velocity fields and distribution of turbulence intensities. However, many workers have investigated the effect of different impeller types

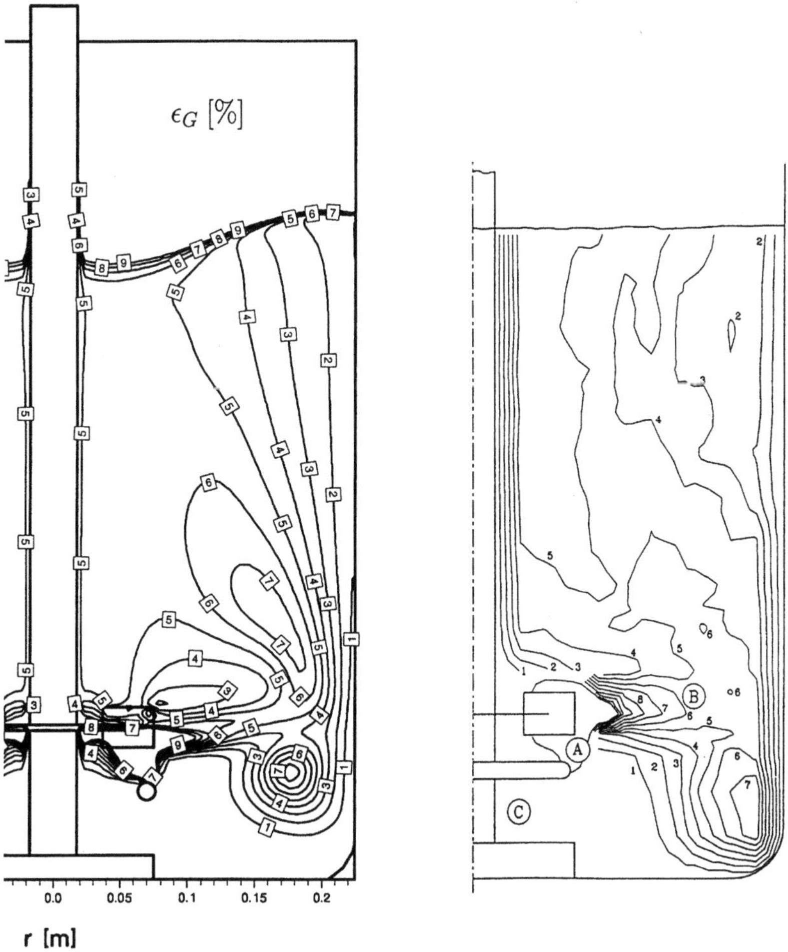

Fig. 5. Comparison between simulated (*left*) and measured (*right*) local gas hold-up at $n = 376$ min^{-1} and $\dot{V}_G = 1.67 \times 10^{-3}$ m^3 s^{-1}. Measurements from Bombac [38]

and configurations on mass transfer gas liquid and mixing. Important and quite useful results have been summarized by Bouafi et al. [45] and John et al. [46]. Improved reactor performance has been observed when incorporating mixed flow systems (e.g., with low impeller acting radially, and the upper impeller axially) in a baffled system [47–49]. In particular, for large-scale fed batch fermentations these configurations should offer advantages because of improved axial mixing. A few CFD simulations together with simulations of mixing behavior will serve to elucidate these phenomena.

To reduce complexity and computation time, two-dimensional simulations have been performed for this comparison. In these simulations the baffles are modeled as a momentum sink in the Reynolds equation for the tangential direction [50]. The assumption that these simplified simulations are able to rea-

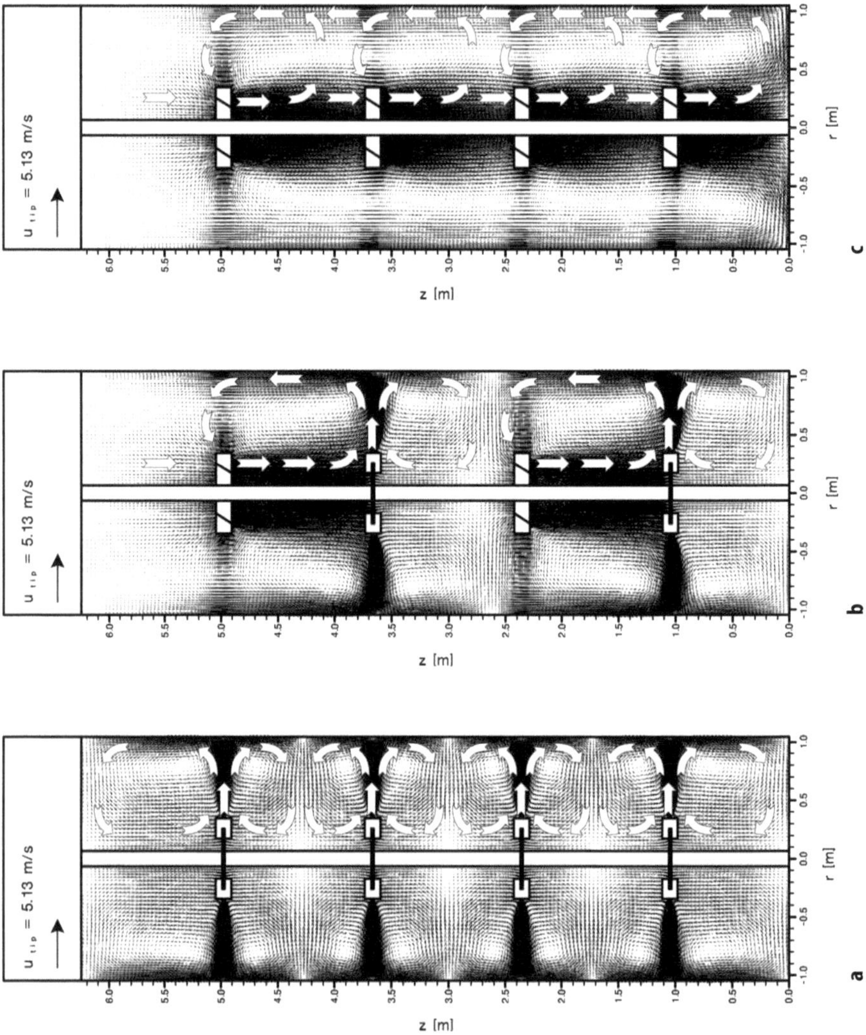

Fig. 6 a–c. Velocity fields at $n = 140\ \text{min}^{-1}$ in different multiple impeller systems. **a** Four Rushton turbines, **b** two Rushton turbines and two pitched blade impellers, and **c** four pitched blade impellers

sonably approximate the radial-axial flow behavior seems to be justified because of the large height to diameter ratio.

In Fig. 6a velocity fields are shown for a system of four Rushton turbines. In addition to the velocity vector field, large arrows are used to illustrate the flow behavior. Each impeller creates a more or less independent symmetrical flow field. The multiple impeller system therefore shows very poor axial convection. The transport between the individual cells is performed mainly with the aid of axial turbulent dispersion.

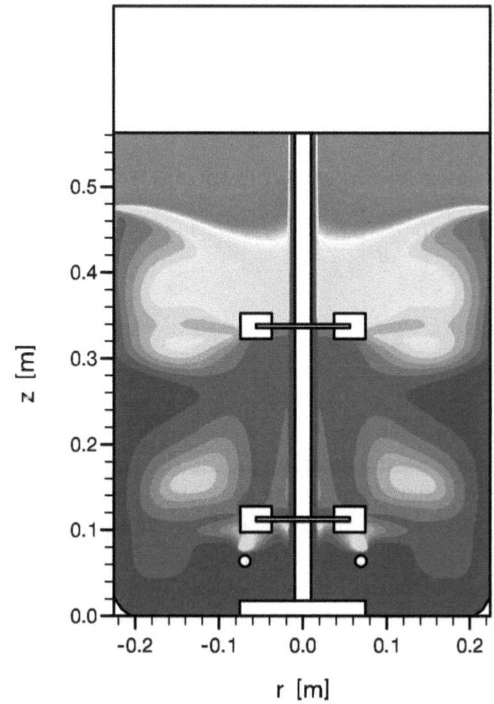

Fig. 7. Local gas hold-up in a stirred tank with double Rushton turbine at $n = 376\ \text{min}^{-1}$ and $\dot{V}_G = 1.67 \times 10^{-3}\ \text{m}^3\ \text{s}^{-1}$. Simulation with two-fluid model

The results from similar simulations with two Rushton turbines and two pitched blade impellers as well as four pitched blade impellers are shown in Figs. 6b and 6c, respectively. In both cases an improved convection can be observed in the axial direction. The results of simulated mixing experiments presented in the following will confirm more rapid mixing for both systems.

In principle it is also possible to extend the simulations for multiple impellers to gassed systems. Figure 7 shows an example of the distribution of the specific gas hold-up for a reactor equipped with two Rushton turbines.

The simulation is based on the three-dimensional Eulerian two-fluid model as discussed for the single impeller. Results are promising as far as the comparison between predictions and measurements of the integral specific gas hold-up is concerned. Routine applications of these simulations are, however, constrained by the tremendous time for computation. A dramatic reduction of this computing time can be achieved with the aid of the so-called algebraic slip model. In this simplified approach the inertial forces are neglected in the momentum equation of the gas phase and only drag is considered as interfacial force. Equation (15) becomes:

$$0 = \frac{\partial p}{\partial x_i} + \frac{3}{4} c_d \frac{\varrho_L}{d_b} |\Delta u| \Delta u_i \qquad (32)$$

The simulation procedure is then simplified and includes the following steps:
(1) Simulation of liquid phase with reduced pumping capacity of the impellers.
(2) Estimation of the representative bubble diameter.
(3) Solution of the simplified momentum balance equation of the gas phase (Eq. (32)).
(4) Solution of a balance equation for the specific gas hold-up:

$$\frac{\partial \varepsilon_G}{\partial t} + u_i^G \frac{\partial \varepsilon_G}{\partial x_i}\left(D_{\mathrm{eff}} \frac{\partial \varepsilon_G}{\partial x_i} \right) + S_G \tag{33}$$

with

$$S_G = \frac{\dot{V}_G}{V_i}$$

at the location of the gas sparger.

There are two main drawbacks of gas-liquid flow simulations based on the algebraic slip approach. Calculation of the gas hold-up with Eq. (33) implies the assumption that the gas phase does not occupy any volume and the gas bubbles do not cause expansion of the volume of the system. Some correction for improving the situation can be performed with the aid of an estimated gas hold-up for the calculation of a reduced mixture density and a volume expansion for the single-phase simulation. Secondly, other important interfacial forces are not considered in the gas phase momentum balance. Neglect of the added mass force especially leads to errors in the simulation of the gas phase, as recognized by Jenne [43] and Friberg [51]. Nevertheless, simulations of gas-liquid flow with the algebraic slip model yield reasonable results. For the tank configuration of Noorman [52] (four Rushton turbines, $V=22$ m^3) a total gas hold-up of $\varepsilon_G=13\%$ has been measured at a stirrer speed of 115 min^{-1} and a gas sparging rate of 0.09 m^3 s^{-1}, while the simulation with algebraic slip results in $\varepsilon_G=11.6\%$.

For these simulations power consumption for the gassed systems was predicted from the empirical correlation for multiple gassed impellers suggested by Cui et al. [53]:

$$\frac{P_G}{P_0} = 1.0 - 9.9\left(\frac{\dot{V}_G}{d_i^2} n^{0.25}\right) \quad \text{for} \quad \frac{\dot{V}_G}{d_i^2} n^{0.25} \le 0.055 \tag{34}$$

$$\frac{P_G}{P_0} = 0.48 - 0.62\left(\frac{\dot{V}_G}{d_i^2} n^{0.25}\right) \quad \text{for} \quad \frac{\dot{V}_G}{d_i^2} n^{0.25} \ge 0.055 \tag{35}$$

3 Coupling of Momentum and Material Balance Equations with Unstructured Biokinetics

The approach used for this application of CFD is illustrated in Fig. 8. It is based on the assumption that the stationary flow field is not affected by mass transfer and reactions. This is a reasonable assumption for many biotechnical processes with Newtonian flow behavior.

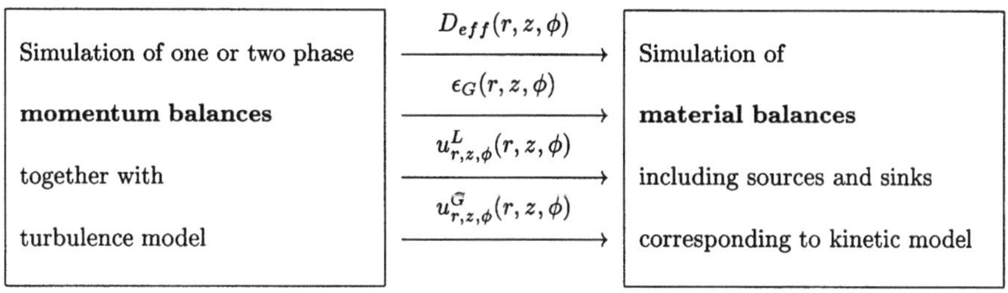

Fig. 8. Separation of momentum and material balance equations

Also, the depletion of oxygen from the gas phase is rather low and usually compensated by the desorption of carbon dioxide. The methodology is attractive because it permits a separation of fluid dynamics (momentum balances, continuity equations, and turbulence model) from material balance equations for the state variables of interest. Figure 8 illustrates how results from the fluid dynamic simulations (mean velocities $u^{G,L}_{r,z,\varphi}(r, z, \varphi)$, turbulent dispersion coefficient $D_{eff}(r, z, \varphi)$, and local gas hold-up $\varepsilon_G(r, z, \varphi)$ can be used as parameters in the material balance equations.

The main advantage of the separation is the reduction of the computational effort. Another aspect is the fact that the two sets of equations can be solved with different numerical methods and on different numerical grids. Due to the nature of the non-linearities in material and momentum balance equations, they usually require different grid refinements in different areas of the computational domain. Additionally, if the assumption of a stationary flow field is valid, the simulation of the coupled set of equations would be unnecessarily slowed down by solving the momentum equations. The material balance to be solved for each of the reacting components k reads

$$\frac{\partial c_k}{\partial t} + u_i \frac{\partial c_k}{\partial c_i} = -\frac{\partial}{\partial x_i}\left(D_{\text{eff}} \frac{\partial c_k}{\partial x_i} \right) + S_k \qquad (36)$$

S_k = reaction + mass transfer gas-liquid + inlets/outlets
$D_{\text{eff}}(r, z, \varphi)$ = turbulent dispersion coefficient = $\upsilon_{\text{eff}}(r, z, \varphi)/Sc_t$

with Sc_t = turbulent Schmidt number, assumed to be one, and $\upsilon_{\text{eff}}(r, z, \varphi)$ from turbulence modeling. Discretization of material balance equations is made using finite volume elements. They are solved either with the differential-algebraic solver Limex (two-dimensional simulations) or ug (unstructured grids, development of the Institute for Computer Applications III, University of Stuttgart, Professor G. Wittum) for three-dimensional simulations.

3.1
Characterization of Mass Distribution via Simulated Mixing Experiments

Investigations of distribution of materials in stirred tank bioreactors are usually based on mixing experiments. For this purpose various methods of pulse injections of tracers have been established and applied to study the intensity of mixing at different operation conditions, vessels varying in volume, number and type of impellers, and geometrical properties. A first step towards coupling the system of material and momentum equations based on CFD is therefore the simulation of these mixing experiments. Pulse experiments can be simulated by solving the material balance equation of the tracer. Figure 9 illustrates an example of the simulated dynamics of a tracer distribution for a single, symmetrically positioned Rushton turbine. The terminal mixing time t_m, used as a quantitative measure, can be easily predicted from the simulated response within the finite volume element corresponding with the sensor position.

From systematic simulations of tracer experiments in a stirred vessel equipped with a Rushton turbine, a height/tank diameter ratio = 1.0, an impeller/tank diameter ratio = 0.3125, and an impeller clearance/height of liquid ratio = 0.31 the following correlation was obtained:

$$n\,t_{m,95} = 27.5$$

With $t_{m,95}$ = time for 95% homogeneity. This result is in reasonable agreement with numbers reported in the literature ($n t_m = 35$ [54] and $n t_m = 32$ [55]).

Despite this good agreement, the terminal mixing time itself does not provide a very informative measure of the mixing process. Because transient concentrations in all volume elements are available from the simulations, it is also possible to calculate the time course of inhomogeneity defined by Landau and Prochazka [56]:

$$I(t) = \frac{1}{V\left(c^{\infty} - c^{0}\right)} \sum_{i=1}^{n} V_i \left|\left(c_i(t) - c^{\infty}\right)\right| \tag{37}$$

Figure 10 summarizes the transient homogeneity for three simulated tracer experiments differing in the position of the tracer input.

The following example serves to illustrate that tracer experiments also help to discriminate between different turbulence models. For this purpose, the classical tracer experiments suggested by Khang and Levenspiel [57] have been employed. In this experiment (also described by Tatterson [58] and Reuss and Bajpai [36]), the pulse injection of the tracer is made with the aid of a concentric ring near the stirrer tips, and the response is measured with a concentric ring electrode nearby. As a simple representation of the recirculation flow for an impeller symmetrically placed in a tank with $H/D_T^{-1} = 1$, Khang and Levenspiel [57] suggested a tank-in-series model with a recirculation loop. The two required parameters, circulation time t_c and number of tanks in the cascade n_t can be predicted from the frequency and decrease of the amplitude of the measured or simulated response, respectively. Figure 11 shows a measured [57] and simulated (material balance Eq. (36) and CFD) response. Simulated and measured responses agree qualitatively well. The more interesting result of this exercise is illustrated in Fig. 12. The simulated response, obviously, is very sensitive to the turbulence model used in the simulations. As a consequence of the overestimation of the turbulent viscosity by the standard k–ε model, the corresponding simulations show

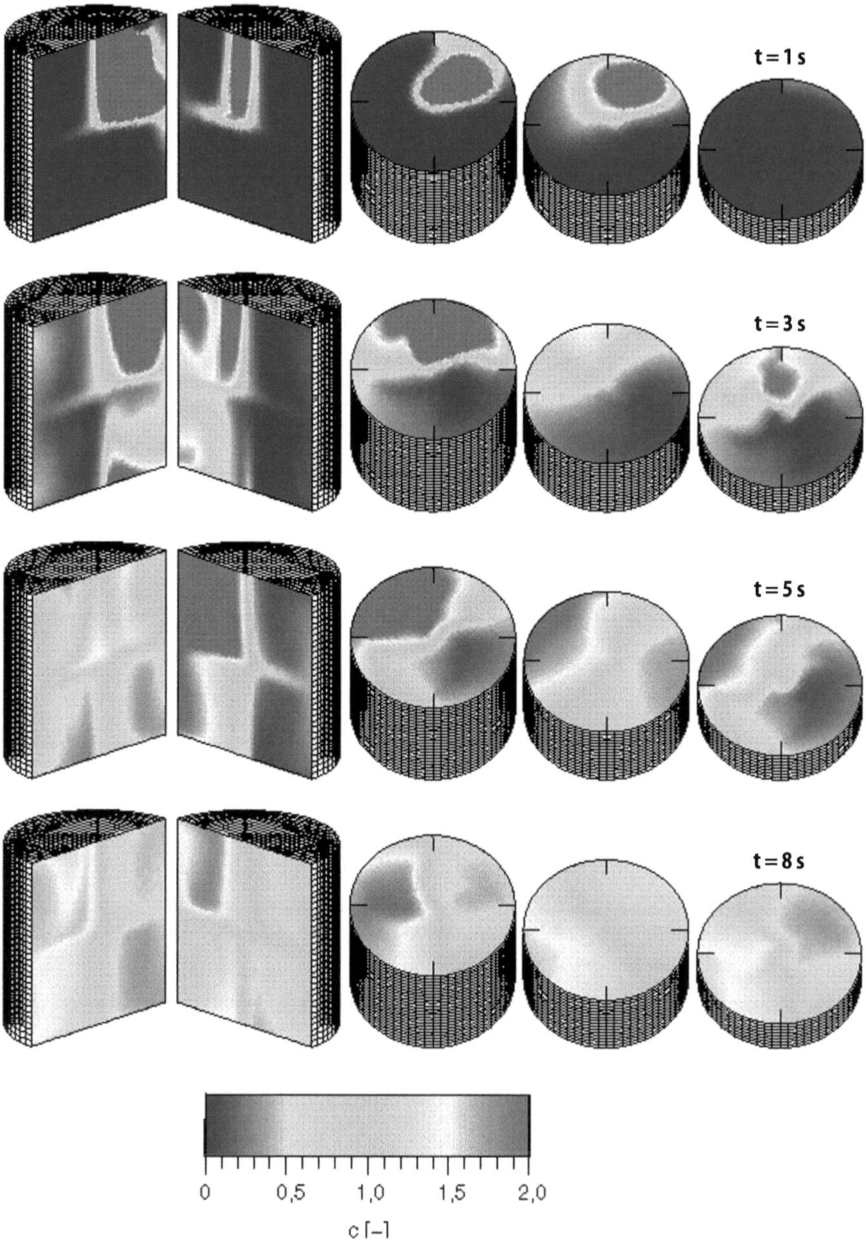

Fig. 9. Tracer distribution at different times after a pulse onto the liquid surface

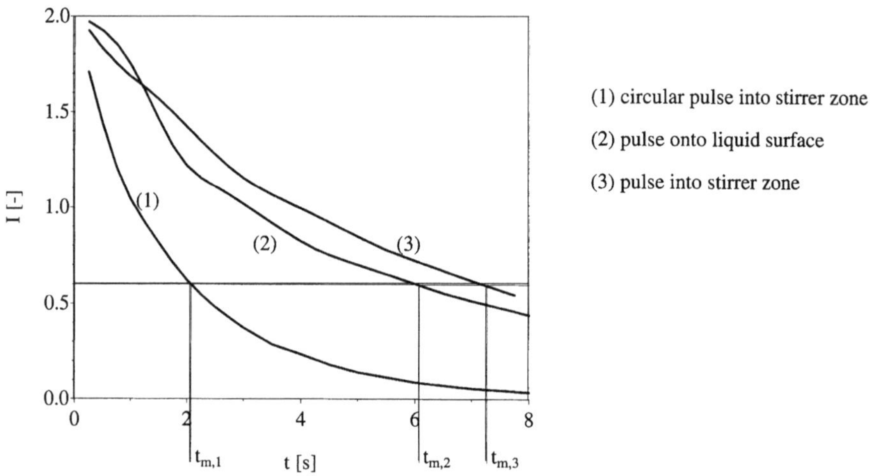

Fig. 10. Inhomogeneity for three different positions of tracer input

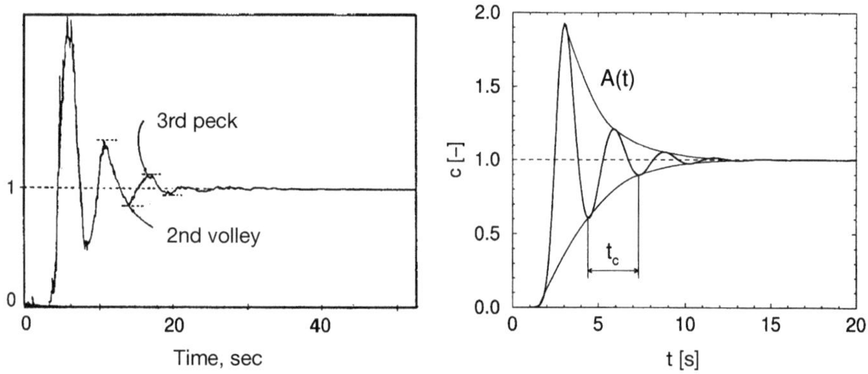

Fig. 11. Measured (*left*, Khang and Levenspiel [57]) and simulated (*right*) tracer pulse response

a response in which, in contrast to the experimental observations, the oscillations are significantly damped down.

Large-scale fermentation equipment usually contains multiple impellers. Pulse experiments for the determination of mixing times have been simulated in a stirred tank configuration with four stages and a liquid volume of 22 m^3. This tank configuration has also been used by Noorman [52], Cui et al. [53], and Friberg [51] in their investigations. The dominating influence on macromixing and mixing times in a multi-impeller system with high *H/DT* ratio is exerted by the axial component of convection. Multiple Rushton turbines are known to cause strong axial flow barriers leading to a compartmentation of the tank volume.

This can be avoided using axial flow impellers such as pitched blade turbines. Figure 13 shows tracer concentrations 60 s after a pulse onto the liquid surface

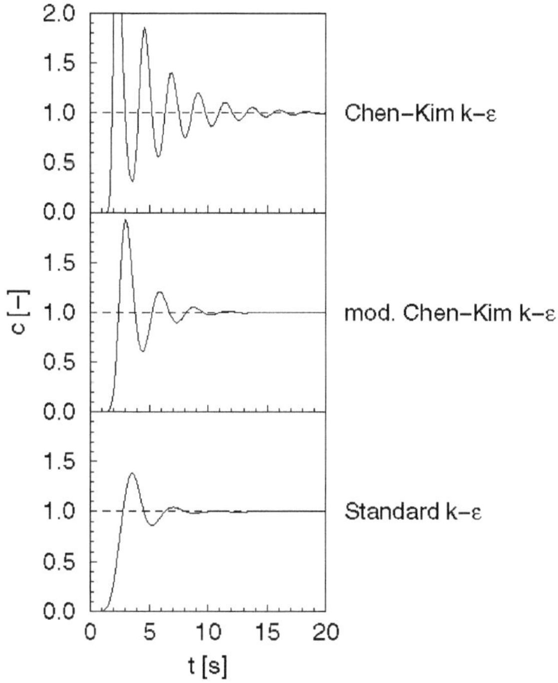

Fig. 12. Influence of turbulence model on simulated tracer pulse response

at a stirrer speed of 140 min^{-1} for stirred tanks with four Rushton and four pitched blade turbines together with profiles of the axial convective flow summed up over the cross section of the tank. With the Rushton turbines, no tracer had reached the lower part of the tank after 60 s. A strong compartmentation is clearly visible. In the tank with pitched blade turbines, complete homogeneity was reached after this time. Simulated mixing times $t_{m,95}$ were 294 s for the Rushton turbines and 18 s for the pitched blade impellers. Mixing time with Rushton turbines is about sixteen times higher than with pitched blade impellers, though this configuration requires only about a third of the power input. It should be noted, that this power input may be necessary for a sufficient oxygen transfer. Also, pitched blade turbines are known to have poor gas disperging capabilities; thus, a combination of Rushton turbine as the gas disperging impeller with pitched blade impellers for good axial macromixing is a better combination.

A comparison of simulated and measured mixing times is given in Table 3. Measurements in the simulated tank have been made by Noorman [52] and Cui et al. [53], while Groen [59] suggests a general correlation for mixing times in multi Rushton-impeller systems based on experiments in stirred tanks up to three stirrers and volumes of 130 m^3. The measurements of Cui et al. and Noorman result in values lower than the simulated times. On the other hand, the correlation of Groen produces a mixing time that is even higher than the simulated values.

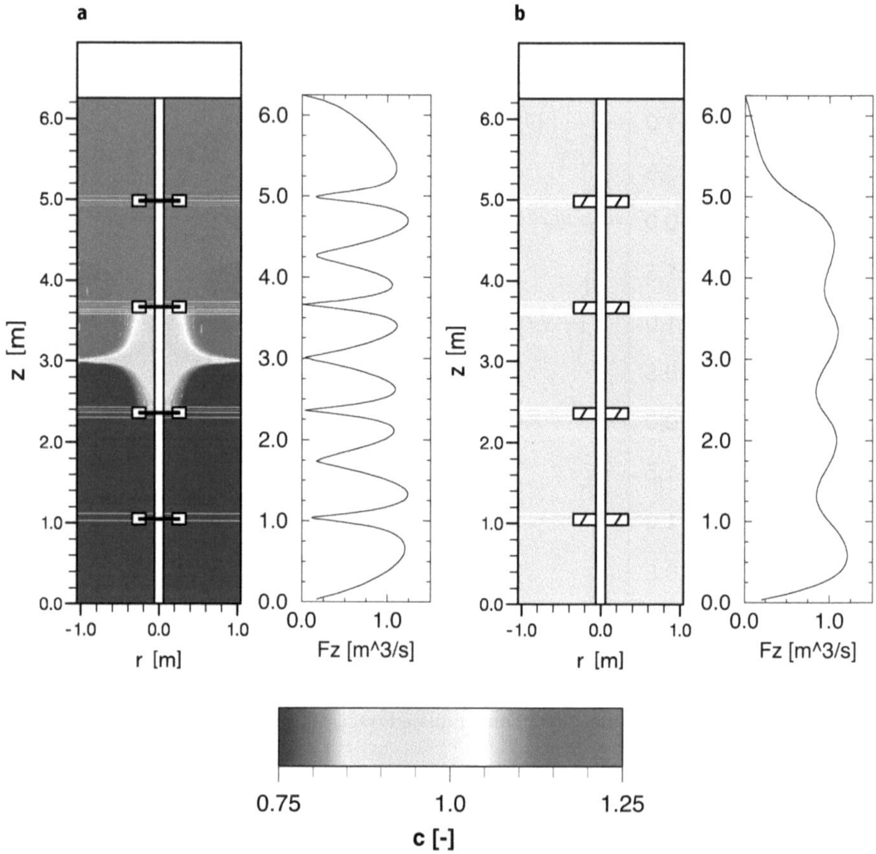

Fig. 13. Tracer concentrations 60 s after a pulse onto the liquid surface at $n = 140\ \text{min}^{-1}$. **a** Four Rushton turbines and **b** four pitched blade impellers together with profiles of the axial convective flow summed up over the cross section of the tank

Table 3. Measured and simulated mixing times in a stirred tank with four Rushton turbines

	Simulation	Noorman [52]	Cui et al. [53]	Groen [59]
Mixing time $t_{m,95}$ (s)	294	139	150	374

A critical assessment on the comparisons leads to the conclusion that agreement between measurements and predictions needs to be further improved for multiple Rushton turbines. In spite of the well-known inaccuracy of measured mixing times, part of these deviations may be also caused by the uncertainty of the turbulence model, the discretization of dispersion, and the dimensionality of the simulation. The comparison between the mixing behavior of multiple Rushton turbines and multiple pitched blade impellers illustrates that the intensity of mixing in these large tanks is mainly determined by the convection in the axial direction. Because axial velocities at the boundary between the different stirrer compartments are very low in the case of Rushton turbines, results of the integral mixing process are very sensitive to uncertainties in the turbulent dispersion across these virtual barriers.

3.2 Simulations of Substrate Distribution in Fed Batch Fermentations

The following example shows how simulations can be applied for optimization of fed batch processes, which are very common for a variety of biotechnical processes to avoid oxygen limitations, heat transfer problems, over-flow metabolism, or catabolite regulation. The concentration of the carbon and energy source in the feed is as high as possible (in the range of 500 kg m^{-3}). The concentration inside the tank very often is in the range of the saturation constant, for example, in the concentration range of 1 – 100 mg L^{-1}. The challenge in the scale up of these processes is then to prevent concentration gradients resulting in further limitations ($c_s < c_{s,crit}$) as well as unwanted byproduct formation or inhibition of production rates ($c_s > c_{s,crit}$). Examples of overflow metabolism are the growth of *Saccharomyces cerevisiae* (production of ethanol) and *Escherichia coli* (production of acetate).

Assuming that oxygen supply is sufficient to avoid local oxygen limitations, the kinetic model required for the simulation includes only the material balance equation for the substrate. As suggested in earlier simulations based on recirculation models (micro-macromixer) by Bajpai and Reuss [60], the uptake kinetics are only considered in the vicinity of the so-called critical sugar concentration. Thus, a rather simple unstructured empirical model is chosen for the purpose of this study. It involves a Monod type of kinetics for substrate uptake

$$S_S = -q_S^{\max} \frac{c_S}{K_S + c_S} c_X \tag{38}$$

which holds at a certain time of the process for the corresponding biomass concentration c_X. If substrate concentration c_S locally exceeds the critical concentration $c_{S,crit}$, an ethanol production rate is superimposed which is given by:

$$S_P = q_P^{\max} \frac{c_S - c_{S,crit}}{K_P + \left(c_{S,crit} - c_S\right)} c_X \tag{39}$$

Equations (38) and (39) are then used as source terms in the material balance equation Eq. (36). Additionally, the feeding rate is considered as a source term in the volume element corresponding to the feeding point. Figure 14 shows results of simulations of the substrate distribution at three different positions for the substrate inlet for a vessel with a volume of 68 L. As expected, feeding of the concentrated sugar solution into the impeller region leads to the best equidistribution of substrate.

Again, simulations have been performed for large-scale vessels with multiple impellers. Figure 15 summarizes the distribution of sugar in a vessel of 22 m^3

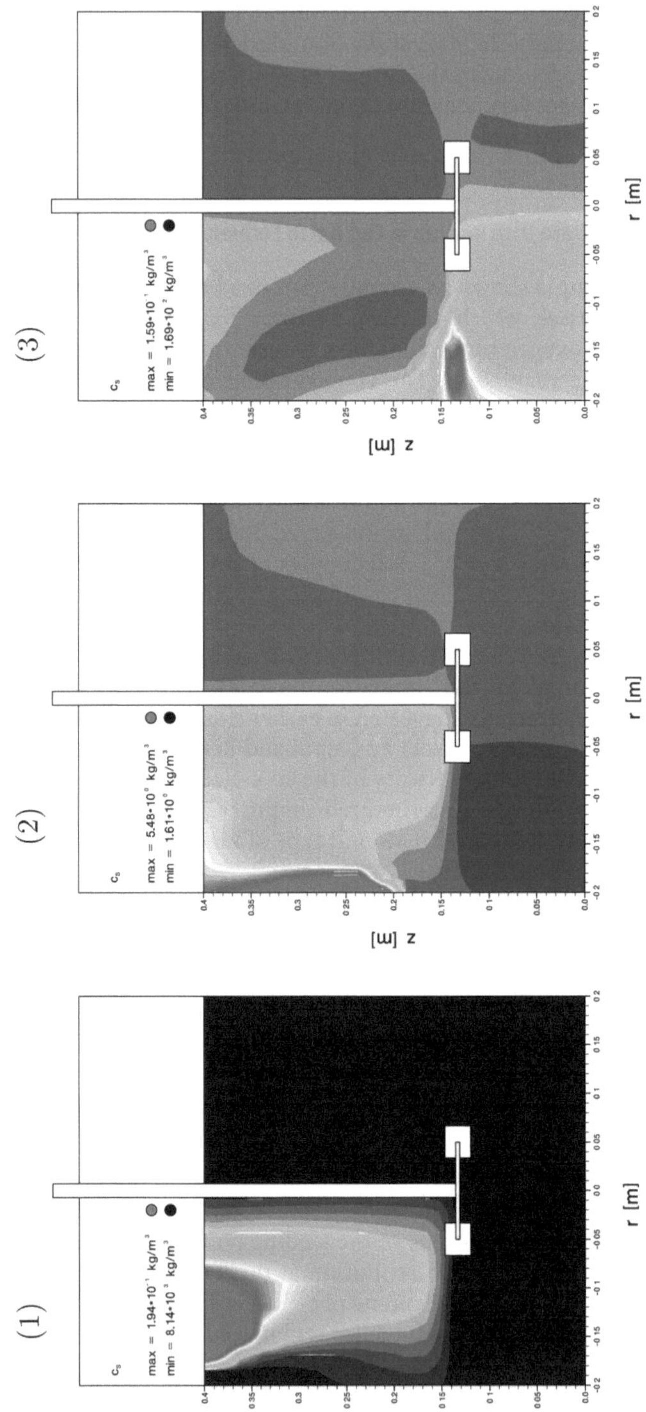

Fig. 14. Substrate distribution for three different feeding positions. *1 Liquid surface*, *2* near tank wall, and *3 stirrer zone*, $V_T = 68$ L

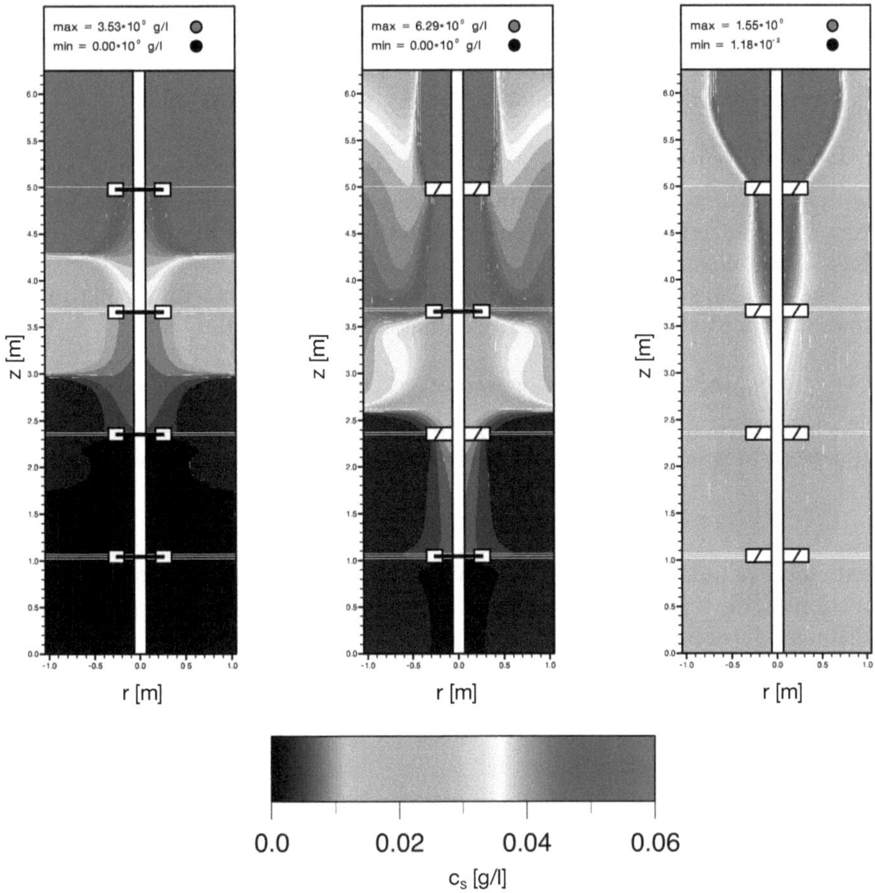

Fig. 15. Substrate distribution for different combinations of Rushton and pitched blade impellers, $V_T = 22\ m^3$

equipped with different combinations of Rushton turbines and axial impellers. As expected, the axial impellers generally lead to a better distribution of the substrate.

It is important to recognize that the configuration of four Rushton turbines not only leads to regions in which the substrate concentrations are higher than the critical value. Particularly in the lower part of the tank, we find regions with pronounced substrate limitations causing further limitations of substrate uptake.

3.3 Distribution of Dissolved Oxygen

An important design consideration of aerobic fermentations is the adequate provision of the oxygen requirements of the culture. This field is almost as old as the art of bioprocess scale up, but remains a major challenge in context with high cell density cultures extensively used in recombinant protein production and processes with metabolically engineered strains for the production of small molecules.

Overall oxygen transfer in bioreactors is given by the expression:

$$S_{O_2} = k_L\, a\, \Delta c_{O_2} \tag{40}$$

with $k_L a$ = overall, averaged volumetric mass transfer coefficient gas-liquid and Δc_{O_2} = mean driving force between concentrations at the gas-liquid interface and bulk of the biosuspension.

The detailed simulations of the fluid dynamic provide us with further information to predict the local intensity of mass transfer according to:

$$S_{O_2}(r,z,\varphi) = k_L(r,z,\varphi)\, a(r,z,\varphi)\left(c^*_{O_2}(r,z,\varphi) - c_{O_2}(r,z,\varphi)\right) \tag{41}$$

The local mass transfer coefficient $k_L(r, z, \varphi)$ is estimated from the correlation suggested by Kawase and Moo Young [61]:

$$k_L(r,z,\varphi) = 0.301\left(\varepsilon(r,z,\varphi)\, v_L\right)^{1/4} Sc^{-1/2} \tag{42}$$

with local values of the energy dissipation rate $\varepsilon(r, z, \varphi)$ available from the simulations of the turbulent two-phase fluid dynamics. Assuming a constant bubble diameter d_b (non-coalescing system) the specific gas liquid interface can be predicted from:

$$a(r,z,\varphi) = \frac{6\varepsilon_G(r,z,\varphi)}{d_b} \tag{43}$$

The concentration at the gas liquid interface in Eq. (41) is calculated from Henry's law:

$$H = \frac{p_{O_2}(r,z,\varphi)}{x_{O_2}(r,z,\varphi)} \tag{44}$$

with mole fraction $x_{O_2}(r, z, \varphi) = c_{O_2}(r, z, \varphi)/\sum c_i$ and partial pressure $p_{O_2}(r, z, \varphi) = y_{O_2}(r, z, \varphi)\, p(r, z, \varphi)$, thus taking into account the local value of the concentration of oxygen in the gas phase as well as the pressure field available from the fluid dynamic simulations.

The kinetics proposed for local oxygen uptake is of a simple irreversible Michaelis-Menten structure which has been verified for the terminal cytochrome oxidase of the respiration chain ($K_{M,O_2} = 1.7\ \mu M$). The distribution of oxygen in the tank is finally computed with the aid of numerical simulations of the oxygen balance equation

$$\frac{\partial c_{O_2}}{\partial t} + u_i \frac{\partial c_{O_2}}{\partial x_i} = \frac{\partial}{\partial x_i}\left(D_{\text{eff}} \frac{\partial c_{O_2}}{\partial x_i}\right) + k_L\, a\left(c^*_{O_2} - c_{O_2}\right) - q^{\max}_{O_2} \frac{c_{O_2}}{K_{M,O_2} + c_{O_2}} c_X \tag{45}$$

coupled to the momentum balance equations as illustrated in Fig. 8.

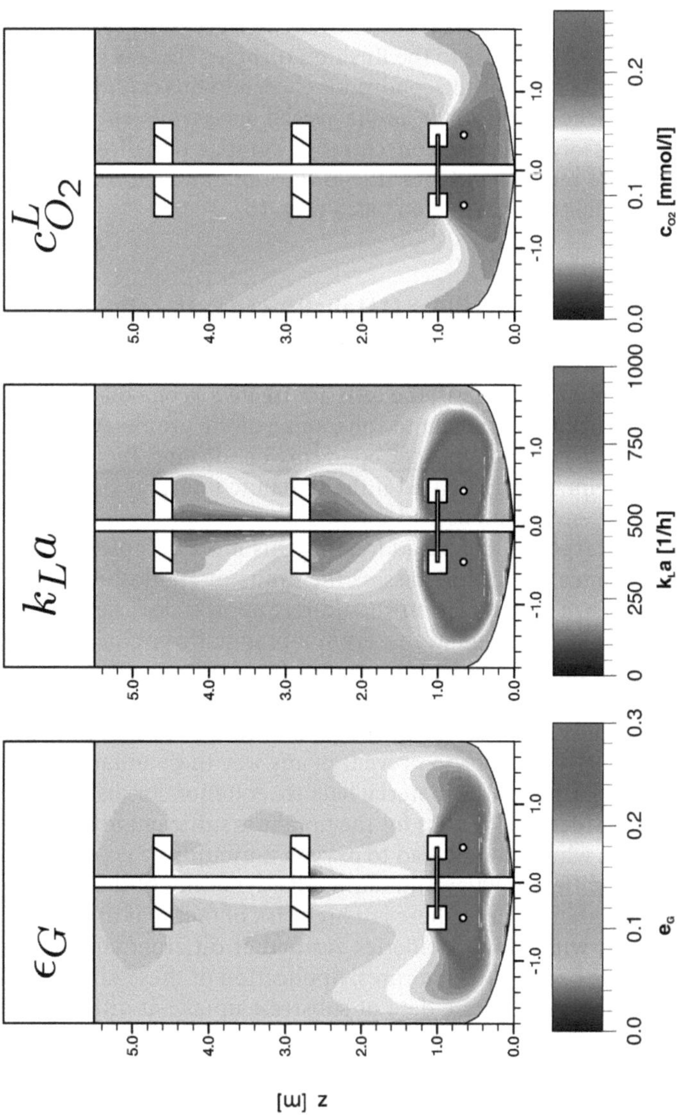

Fig. 16. Local gas hold-up, volumetric mass transfer coefficient, and dissolved oxygen concentration in a stirred tank with one Rushton turbine and two pitched blade impellers. $n=100\ \text{min}^{-1}$, $\dot{V}_G=0.224\ \text{m}^3\ \text{s}^{-1}$, $V_T=54\ \text{m}^3$, $Q_{O_2}^{\max}=\text{mol}\ (\text{m}^3\ \text{h})^{-1}$

Simulated distributions of gas hold-up, mass transfer coefficient, and dissolved oxygen concentrations in a mixed impeller system (*1 Rushton turbine*, 2 pitched blade impellers) are shown in Fig. 16. At a stirrer speed of 100 min^{-1} and a gassing rate of 0.224 $\text{m}^3\ \text{s}^{-1}$, the simulated volume averaged values of gas hold-up and volumetric mass transfer coefficient are $\varepsilon_G=0.17$ and $k_L a=532\ \text{h}^{-1}$, respectively. The value of the volumetric mass transfer coefficient shows a reasonable agreement with $k_L a=610\ \text{h}^{-1}$ estimated from the empirical equation suggested by van't Riet [62]. As illustrated in Fig. 16, the gas hold-up is maximal close to the ring sparger, and the highest values of the mass transfer coefficient are observed in the region of high energy dissipation close to the Rushton turbine. There is no oxygen limitation throughout the tank for the conditions chosen in this example. The mixed impeller system reaches almost 90% of the overall mass transfer coefficient and 80% of the mean dissolved oxygen concentration compared to three Rushton turbines, which require more than twice the power input and show non appropriate mixing of the substrate during fed batch (Fig. 15).

4 Dynamic Response of Intracellular Metabolites to Extracellular Stimuli

The approach presented so far is based upon simple unstructured kinetic expressions of the Monod type; thus, the cells are treated as one black box model. Although this approach is sufficient to solve some of the problems in the design of bioreactors and selection of proper operating conditions, the application of this modest approximation is limited if the dynamic interplay between abrupt changes in the cell environment and intracellular machinery leads to a more complex dynamic response. Before addressing this more complex issue, it is of overriding importance to carefully assess the limits of applicability of the unstructured approach. This evaluation should rest upon a clear definition at the outset, the purpose of the simulations. From a pragmatic engineering point of view it is not a distinguished endeavor to simulate the adventures of cells traveling between aerobic, semianaerobic, and anaerobic conditions within the bioreactor. In contrast, the clear task for the engineer must be to design the reactor and/or the operating conditions to prevent in any way these unfavorable conditions. As such, simulations of the material balance equation for dissolved oxygen based on Michaelis-Menten kinetics for the uptake is sufficient to find those operations conditions which do not lead to oxygen limitations.

The situation is different when simulating the dynamics of the uptake of the carbon and energy source. Here, there is a high risk of failure if the dynamic behavior is predicted with Monod kinetics verified at different snapshot steady states in continuous or fed batch cultures. Application of these kinetics is questionable, because the steady state data of substrate uptake at different dilution rates may be corrupted by induction of different transporter systems depending on the steady state substrate concentrations. In addition to the variability of the affinity of the various transporter systems as clearly demonstrated for the yeast *Saccharomyces cerevisiae*, we do expect pronounced differences between permeases and phospho-transferase systems because of the clear distinctions in the influence of intracellular metabolites upon the uptake dynamics.

The impact of the complex phenomena cannot be evaluated without addressing the issue of intracellular response to fast variations in the extracellular substrate concentrations. The following discussion is therefore targeted as an introduction into this territory of in vivo analysis of the fast dynamics of the cascade of intracellular reactions.

The stimulus-response strategy for the in vivo diagnosis of intracellular reactions uses experimental observations of intracellular metabolites under transient conditions. For this purpose, a continuous culture (or fed batch process at controlled specific growth rate) is disturbed by a pulse of glucose, and changes in metabolite concentrations are measured within seconds or even milliseconds. Precise measurements of intracellular concentrations in these time spans requires appropriate techniques for rapid sampling, inactivation of metabolic reactions (quenching), and extraction of metabolites, taking into account the high metabolic turnover rates of the compounds of interest.

Figures 17 and 18 illustrate two different techniques developed for the aforementioned rapid sampling and quenching. Both sampling devices are connected with a stirred tank bioreactor operating in a continuous mode. In the first approach [63–66], a pulse of glucose is injected into the bioreactor with a syringe to give an initial glucose concentration of for example 1 g L^{-1} (steady state concentration of glucose before the pulse is about 20 mg L^{-1}). Samples are then rapidly taken aseptically with vacuum-sealed, pre-cooled glass tubes through a special sampling device [63–65]. The frequency of sampling is indicated in Fig. 17. The sample tubes contain an appropriate quenching fluid depending on

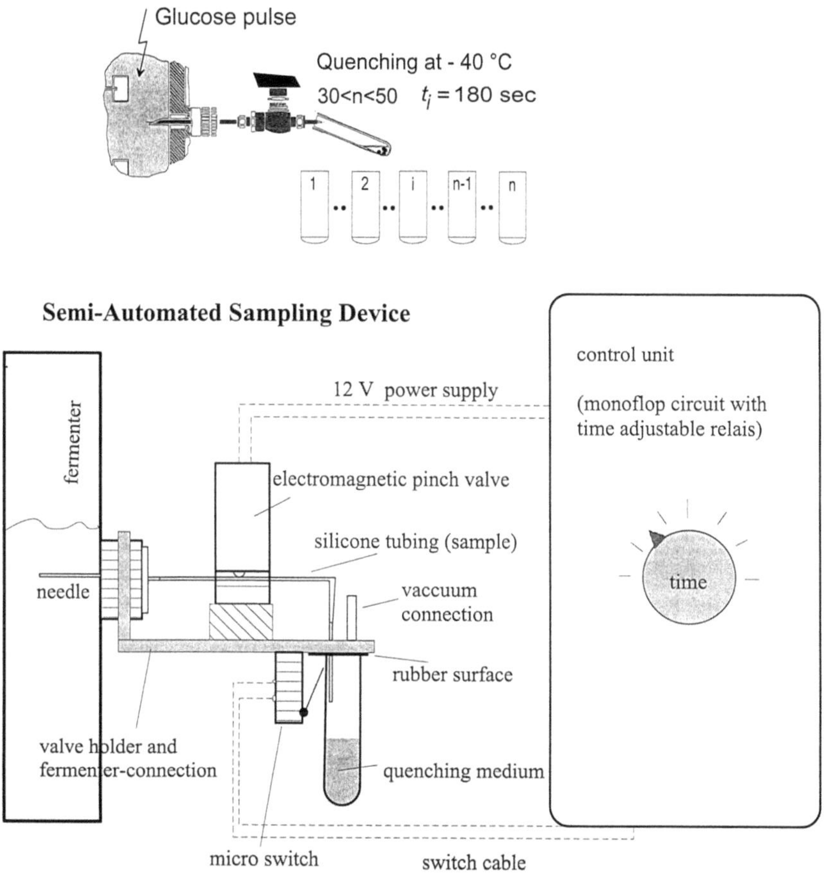

Fig. 17. Rapid sampling and quenching devices

the microorganism and the metabolite to be measured (perchloric acid solution: –20 °C; methanol: –70 °C; liquid nitrogen: –196 °C). Systematic investigations have indicated that the most important quenching effect is due to the low temperature.

This sampling technique can be easily automated to increase the frequency of sampling [67, and Reuss et al., unpublished results]. However, as far as the very fast and initial response of intracellular metabolites in the millisecond range is concerned – and this is the time span of interest for the dynamic situation in the bioreactor – this method shows an inherent limitation. The time span for the first sample after disturbance is determined by the mixing time of the glucose pushed into the bioreactor. Even in small laboratory reactors, mixing times are in the order of 2–3 s.

Figure 18 shows an alternative sampling technique designed to overcome this problem [68]. It is based on the well-known stopped-flow method used for fast measurements of enzymatic reactions. In its application to sampling from bioreactors, a continuous stream of biosuspension leaving the bioreactor is mixed with

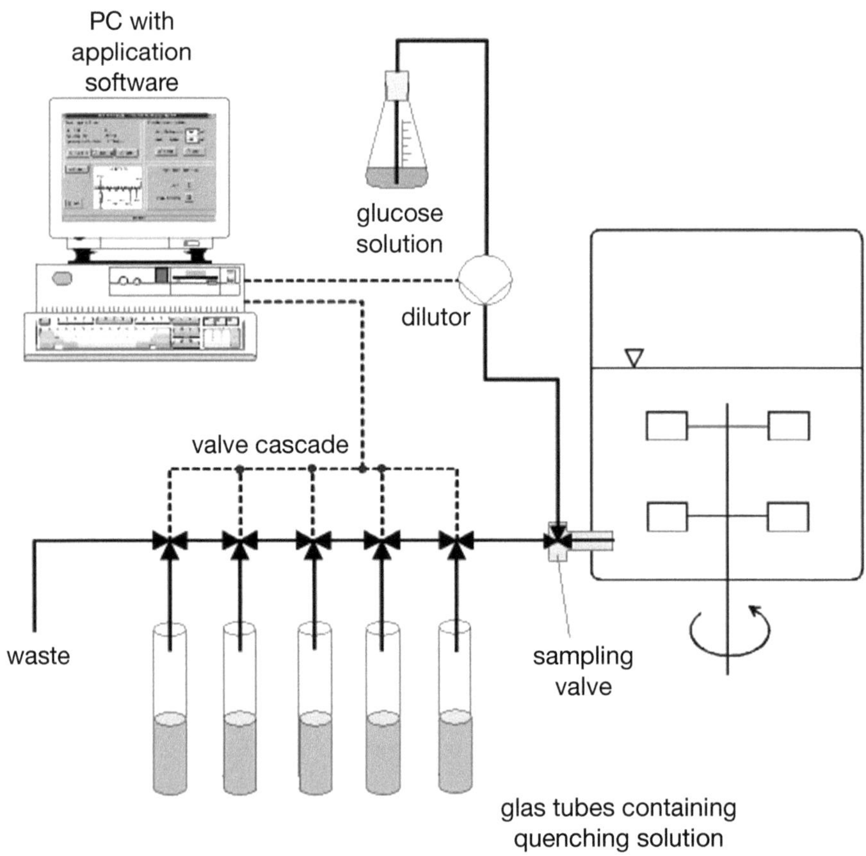

Fig. 18. Stopped-flow sampling technique

a concentrated glucose solution in a turbulent mixing chamber (mixing time: few milliseconds). The position of the valves in the cascade illustrated in Fig. 18 then determines the residence time of the biosuspension before being quenched in the corresponding sampling tube. The main features of this sampling device may be characterized as follows:

(1) Very sharp stimuli, easy to be extended to temperature, pH etc.
(2) The culture remains at steady state because the microorganisms are stimulated by the glucose in the mixing chamber within the valve.
(3) The sampling time and reaction time are decoupled. The volume of the individual samples can be chosen independently.
(4) The time span between glucose stimulus and first sample can be less than 100 ms.

The only limitation of the technique is the problem of oxygen limitation at aerobic growth. Thus, the longest reaction time is determined by the oxygen consumption rate in the sampling tube. For studying the complete response it is therefore recommended to use both sampling devices, the stopped-flow technique for the first seconds, and the pulse technique with manual or automated sampling for longer time periods.

As far as further details and results of dynamic measurements with the manual sampling technique is concerned, the reader is advised to study the original publications for baker's yeast *Saccharomyces cerevisiae* growing under aerobic conditions [63–66, 69, 70]. For the simulation studies of coupled fluid dynamics and intracellular network kinetics illustrated in the next section, an alternative physiological state of the yeast has been used. Figure 19 summarizes the intracellular response of the yeast (*S. cerevisiae* VW1) after a pulse of glucose growing under anaerobic conditions. Measured data are shown along with results of the dynamic simulations based on a structured metabolic model [71]. Figure 20 depicts the topology of the dynamic model.

To describe the dynamic system behavior, deterministic kinetic rate equations of the form

$$r_i = r_{\max,i}\, f(\mathbf{c}, \mathbf{p}) \tag{46}$$

have been formulated, where the maximal rate $r_{\max}$ is obtained from the vector of model parameters $\mathbf{p}$, the vector comprising metabolite and cometabolite and effector concentrations $\mathbf{c}$, and the flux distribution at a specific growth rate of $\mu = 0.1\ \mathrm{h}^{-1}$ (see Fig. 21); accordingly

$$r_{\max,i} = \frac{1}{r_i^{\text{steady state}}}\, f\left(\mathbf{c}^{\text{steady state}}, \mathbf{p}\right) \tag{47}$$

The metabolome's response due to dynamic system excitation has been used to identify the structure of the kinetic expression as well as the model parameters by a stepwise strategy similar to the method proposed by Rizzi et al. [70] and Vaseghi et al. [66]. Table 4 gives further information concerning the structure of the kinetic expressions identified according to this procedure [Mauch et al., to be published].

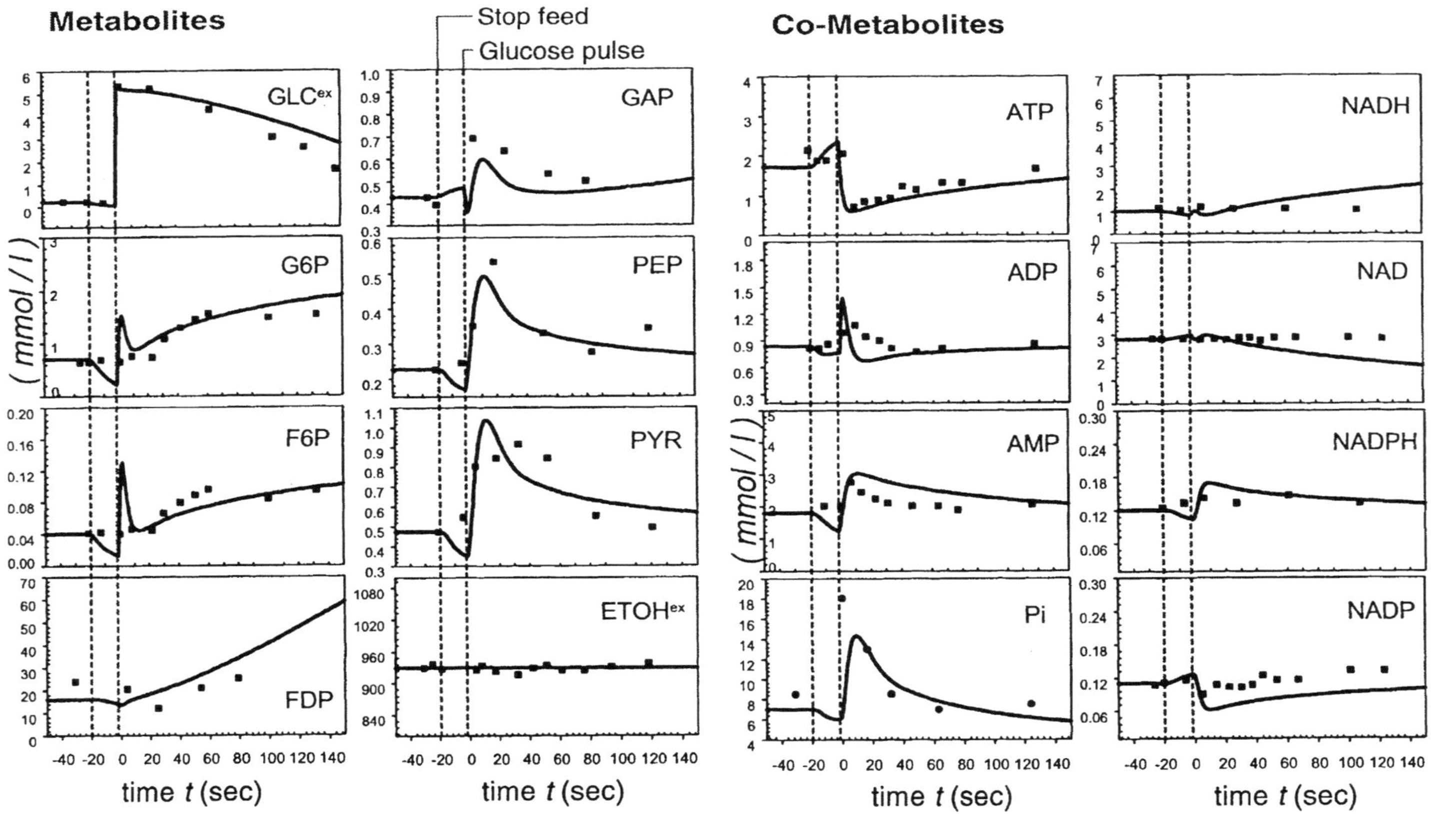

Fig. 19. Comparison between model simulation (*straight line*) and measured concentrations of glycolytic metabolites and cometabolites after dynamic system excitation

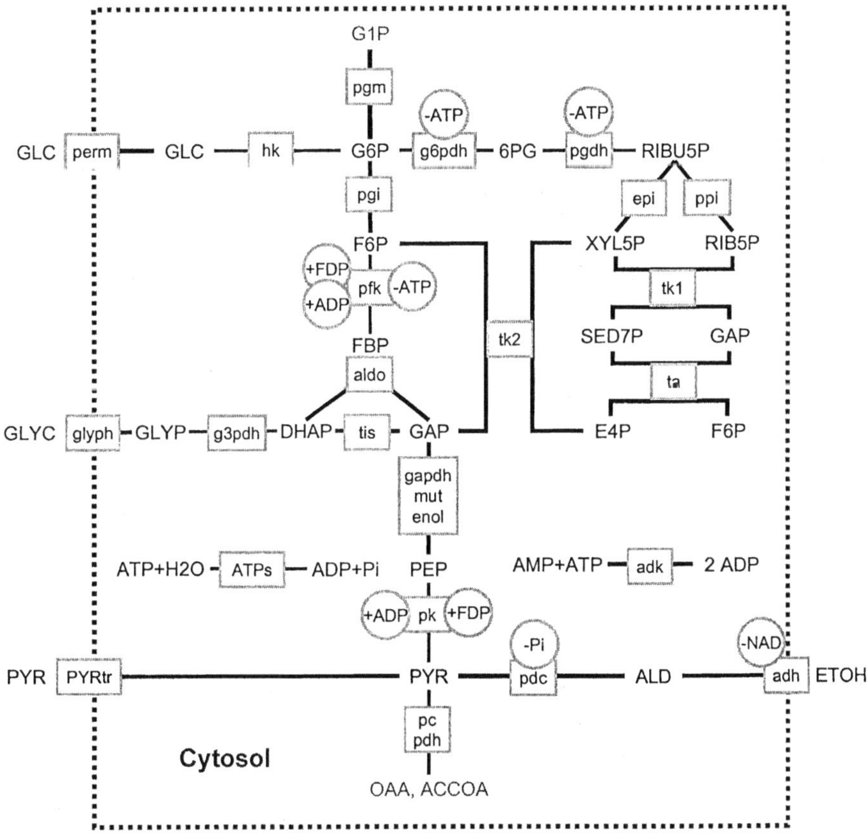

Fig. 20. Topology of the structured metabolic model of the yeast *S. cerevisiae* under anaerobic conditions. Effectors are shown in *circles*

In attempting to simulate the dynamics of the intracellular machinery in response to the concentration fields in the bioreactor, it is worthwhile to get a further insight into the nature of the response within the first few milliseconds. The aforementioned stopped-flow sampling technique should provide us with information about this more rapid transient behavior, and the two examples presented in the following serve to strengthen the relevance of this time span.

Figure 22 shows measurements of intracellular glucose-6-phosphate, AMP, ADP, and ATP concentrations in *Saccharomyces cerevisae* growing under aerobic conditions in a continuous culture [68].

The experimental data have been obtained after a pulse of glucose (time span of seconds) and with the aid of the abovementioned stopped-flow sampling technique (time span of milliseconds). For further details regarding the performance of these measurements the reader is referred to the original publications [63, 65, 68, 69]. The measurements strongly indicate the fast response of the glucose transporter and the subsequent phosphorylation, and confirm the earlier obser-

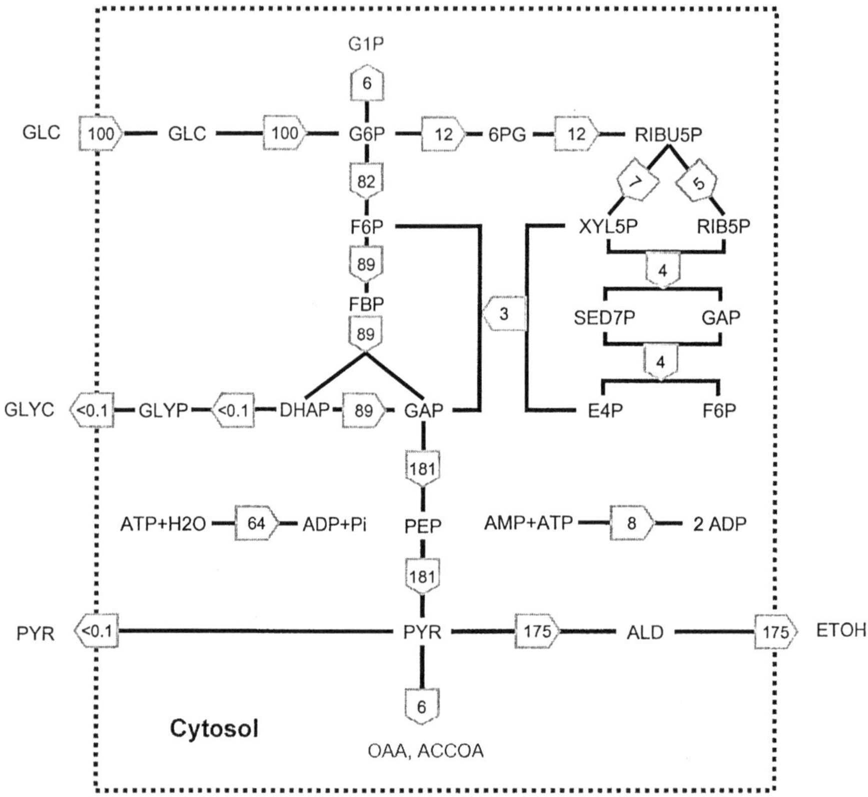

Fig. 21. Flux distribution within central metabolic pathways of *S. cerevisieae* at a mean specific growth rate of $\mu = 0.1\ h^{-1}$ (anaerobic conditions)

vations of de Koning and van Dam [72]. The dynamic response pattern observed with the stopped-flow measurement technique is in line with the manual sampling technique. This circumstance is a first hint that the simulations based on the coupling of the fluid dynamics with the unstructured kinetic rate expression for the uptake system is a reasonable approach for the yeast. Thus, at least the concentration fields for the carbon and energy source are credible and provide a unique resource for predicting the corresponding operating conditions of the bioreactor.

However, the situations may become more sophisticated if uptake systems with stronger links to the nonlinear network dynamics are considered. The important example of the phosphoenol-dependendent phosphotransferase system (PTS) in *Escherichia coli* is discussed next to illustrate this level of complexity. The reaction scheme shown in Fig. 23 is responsible for the concomitant translocation and phosphorylation of several sugars across the cytoplasmatic membrane. Sugar phosphorylation and translocation of glucose appears to involve several phosphoproteins, intermediates, and phosphoryl transfer reactions.

Table 4. Kinetic expressions in the model of *Saccharomyces cerevisiae*

Name	Symbol	Mechanism	Activator	Inhibitor
Alcohol dehydrogenase	adh	Competitive		NAD
Adenylate kinase	adk	Near equilibrium		
Aldolase	aldo	Ordered uni bi		
ATP consumption for maintenance	ATPs	Michaelis-Menten		
Epimerase	epi	Near equilibrium		
Glycerol 3-phosphate dehydrogenase	g3pdh	Ordered bi-bi		
Glucose 6-phosphate dehydrogenase	g6pdh	Competitive		ATP
6-Phosphogluconate dehydrogenase	pgdh	Competitive		ATP
Glyceraldehyde 3-phosphate dehydrogenase	gapdh mutenol	Michaelis-Menten, reversible		
Phosphoglycerate mutase Enolase	enol			
Glycerol phosphatase	glyph	Michaelis-Menten		
Hexokinase	hk	Rapid equilibrium random bi-bi		
Pyruvate carboxylase	pc	Michaelis-Menten		
Pyruvate dehydrogenase	pdh			
Pyruvate decarboxylase	pdc	Allosteric		Pi
Permease	perm	Michaelis-Menten, reversible		
Phosphofructo 1-kinase	pfk	Allosteric	ADP, AMP	ATP
Phosphoglucose isomerase	pgi	Michaelis-Menten, reversible		
Phosphoglucose mutase	pgm	Michaelis-Menten		
Pyruavate kinase	pk	Allosteric	FDP, ADP	ATP
Triosephosphate isomerase	tis	Michaelis-Menten, reversible		
Transketolase	tk	Near equilibrium		
Pyruvate transport	TRpyr	Michaelis-Menten		

Elucidating the dynamics of this uptake system is an essential step to critically assess the reliability of approximations based on the unstructured Monod type of rate expression. To address this issue, we applied the tools of measurements of metabolite concentrations and in vivo diagnosis to *E. coli* (Noissomit-Rizzi et al., to be published). Both sampling techniques were used during continuous culture of *Escherichia coli* W3110. In what follows, only those metabolites will be discussed which are related to the uptake system. Also, a first attempt towards kinetic analysis of the PTS system will be presented.

Figure 24 shows the extracellular glucose concentration after the continuous culture has been disturbed by the pulse of glucose. At a first glance, this part of the dynamic response is neither surprising nor really interesting. What can be observed is a remarkable increase of flux, which results in acetate excretion (not shown) after a short time delay. There are, however, clear signs that the initial response points to a delay in the abrupt increase of the uptake. This behavior would have enormous implications in the dynamic behavior of the cells traveling through regions of varying concentrations within the bioreactor.

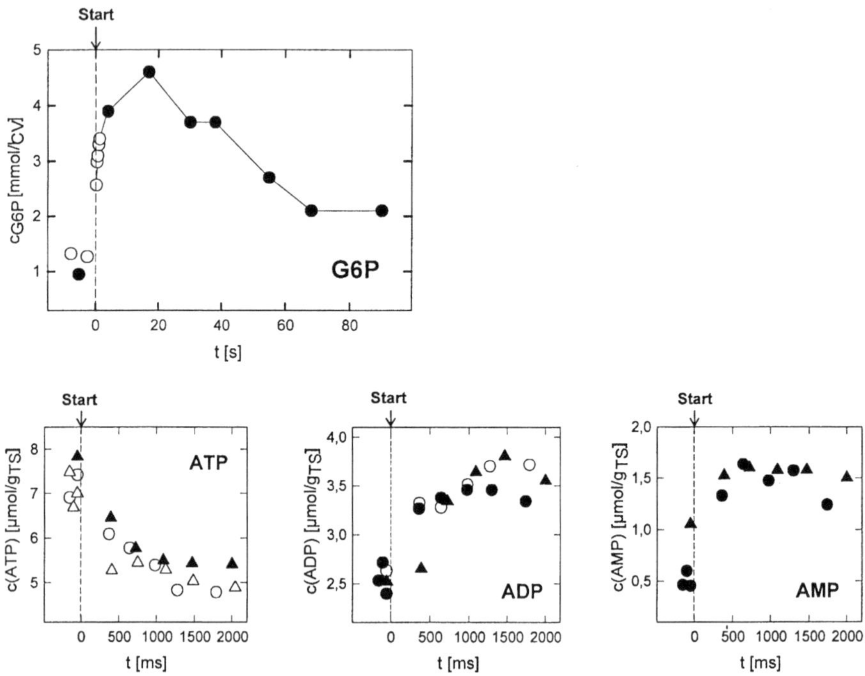

Fig. 22. Intracellular response of glucose-6-phosphate (G6P), AMP, ADP, and ATP *in S. cerevisiae* in response to a glucose stimulus in continuous culture ($D = 0.1\ h^{-1}$) measured with the aid of the stopped-flow sampling technique. Long-term glucose-6-phosphate measurements from Theobald et al. [65]

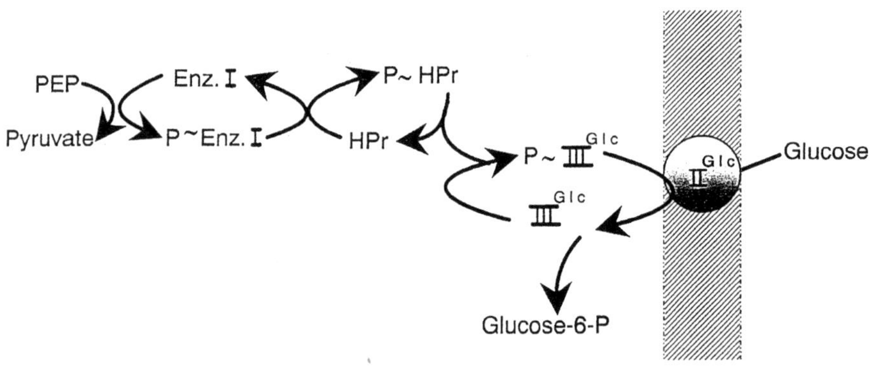

Fig. 23. Reaction scheme of the phosphoenol-dependent phosphotransferase system (PTS) in *Escherichia coli*

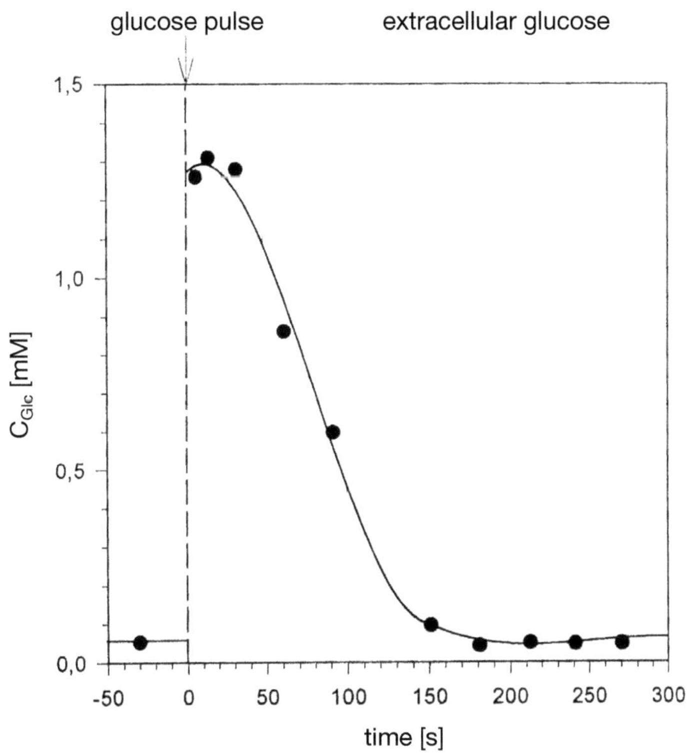

Fig. 24. Transient glucose concentration in response to a glucose pulse during continuous cultivation of *Escherichia coli* ($D = 0.1\ h^{-1}$)

The interpretation of the more interesting intracellular response will be based on a simple kinetic model for the PTS system. The rate expression suggested by Liao et al. [73] can be derived from the reaction scheme in Fig. 23 assuming that the three phosphoryl transfer reactions

$$PEP + E_I = E_I - P + PYR$$
$$E_I - P + HPr = HPr - P + E_I$$
$$HPr - P + E_{II} = E_{II} - P + HPr$$

are in equilibrium and reads:

$$r_{Gluc} = \frac{c_{Gluc}^{ex}}{K_1 + K_2 \dfrac{c_{PEP}^{in}}{c_{PYR}^{in}} + K_3\, c_{Gluc}^{ex} + c_{Gluc}^{ex} \dfrac{c_{PEP}^{in}}{c_{PYR}^{in}}} \tag{48}$$

According to this model the ratio of the concentrations of PEP to PYR and external glucose concentrations determine the uptake of glucose. Figure 25 shows the results of the measured ratio of the two metabolites in the time span of seconds and milliseconds.

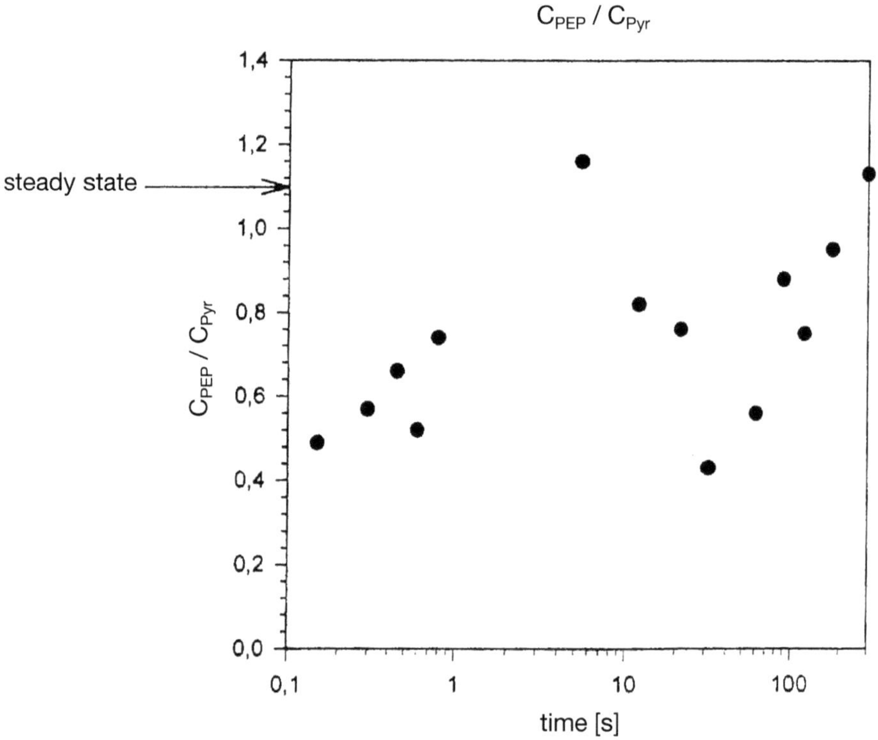

Fig. 25. Intracellular response of the ratio of phosphoenol pyruvate and pyruvate measured after stimulus with glucose. Stopped-flow sampling technique, continuous culture of *Escherichia coli* ($D = 0.1\ h^{-1}$)

Obviously, there is a remarkable drop in the ratio of the two concentrations in the beginning followed by an interesting and complex dynamic pattern over a time span of approximately 3 min. A similar dynamic behavior can be extracted from the data of intracellular pool concentrations presented by Schäfer et al. [74].

The results from a first attempt to use the data along with the rate expressions (Eq. (48)) are summarized in Fig. 26 [84].

According to these predictions, the uptake rate is limited by glucose at steady state. Immediately after the increase of glucose, the uptake increases for a short time span. In the next moment the uptake rate is limited by the ratio of intracellular concentrations of PEP and PYR. The dynamics of this ratio is the result from the superposition of the uptake system and the dynamics of glycolysis. These effects are further amplified by an inhibition of the uptake system through the increasing concentration of glucose-6-phosphate [75].

Besides the interesting dynamic structure of complex intracellular oscillations these results have an important impact on the task of coupling fluid dynamics and cell metabolism. The strong feedback-link between pool concentrations and uptake system leads to a complex system behavior driven by the dynamic

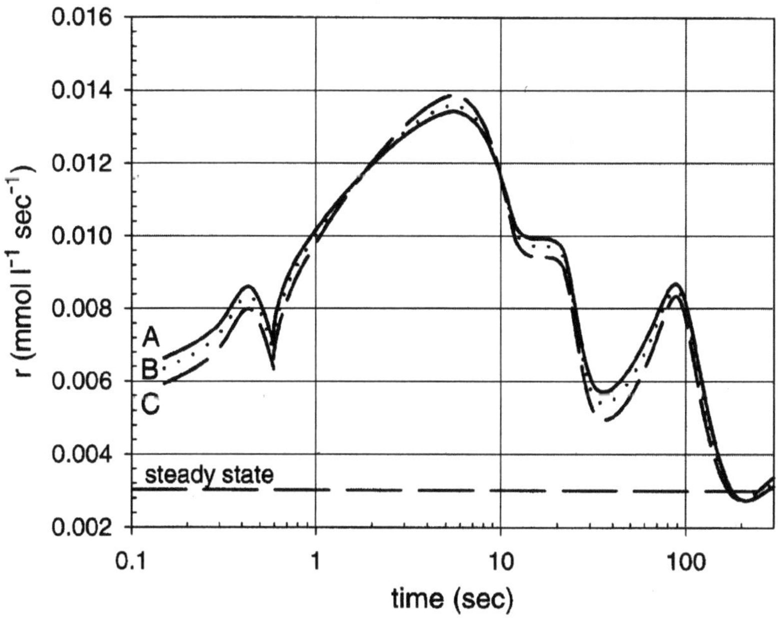

	a_1 (mmol / l)	a_2 (mmol / l)	a_3 (-)	r_{max} (mmol / (l sec))
A	0.24	1	6	0.095
B	0.94	1	10	0.15
C	16.9	1	100	1.37

Fig. 26. Transient behavior of glucose uptake rate calculated from Eq. (48) with the measured ratio of PEP/PYR (Fig. 25) and estimated kinetic parameters

changes in the environment. Because decomposition of this nonlinear behavior is not possible, a more reliable approach of coupling fluid dynamics and physiology should rest on a structured model for the glycolysis of *E. coli* (Chassagnole et al., to be published). In what follows, an introduction into this complex territory will be presented for the yeast *Saccharomyces cerevisiae.*

5 Metabolically Structured Models Stimulated by Dynamically Changing Environment – Integration of CFD and Structured Kinetic Models

To illustrate application of the more complex systems approach of integrating computational fluid dynamics (CFD) with intracellular kinetics we again take as a first example *Sacchararomyces cerevisiae* for which a dynamic model for the glycolysis based upon measurements of intracellular metabolites has been presented earlier [64, 70]. To reduce the complexity of this model, the simplified version for anaerobic growth will be used. Measured data and model structure have been discussed above [71]. Simulations have been performed for a production scale mixed impeller system (see Fig. 16). The stirrer speed is 80 min^{-1} in a tank volume of 54 m^3. The balance equations of 15 intracellular metabolites and the extracellular concentrations of glucose and ethanol have been solved dynamically on an axisymmetric flow field. The glucose pulse has been added onto the liquid surface of a continuous culture of *S. cerevisiae* at a dilution rate of 0.1 h^{-1}. Figure 27 shows the simulated concentration distributions of glucose, ATP, and pyruvate four seconds after the pulse.

The dynamic response of the intracellular pool concentrations driven by the extracellular glucose concentration field is profound. The cells exposed to the spatially inhomogeneous environment obviously never see a steady state [76]. Unraveling the implications of these variations still remains a pivotal problem for future research. For the time being, it is not meaningful to speculate further about this issue in context with the complex metabolism of *Saccharomyces cerevisae* and the difficulties of experimentally verifying the spatial variations of intracellular properties.

However, to convince that the approach is already of practical value we need a system in which exposure to spatial variations of the extracellular environment results in a measurable metabolic response. *Bacillus subtilis* has been used several times as an appropriate model system for oxygen sensitive metabolism to characterize the effects of inhomogeneities in the intensity of oxygen transfer gas-liquid in stirred tank bioreactors [77–79]. At dissolved oxygen concentrations below 1% saturation, the ratio of the two production rates of acetoin and butanediol strongly depends on the dissolved oxygen concentration. In other words, the selectivity of the process is very sensitive to changes in dissolved oxygen under microaerobic conditions. A detailed simulation of these effects requires a model for the oxygen gradients and production rates for the two products related to the dissolved oxygen concentration.

For the simulations presented in the following, a kinetic model proposed by Moes et al. [78, 79] has been used. The model takes into account the formation of biomass, the two products acetoin and butanediol, the substrate and oxygen consumption. The source terms in the material balance equations for the six state variables are given by:

biomass

$$S_X = r_{ATP \to X} Y_{X/ATP} c_X \tag{49}$$

substrate

$$S_S = -\frac{r_{S \to ATP}}{Y_{P/S}} c_X \tag{50}$$

acetoin

$$S_{Ac} = \left(r_{S \to Ac} - r_{AC \to Bu} + r_{Bu \to Ac} \right) c_X \tag{51}$$

butanediol

$$S_{Bu} = \left(r_{AC \to Bu} - r_{Bu \to Ac} \right) c_X \tag{52}$$

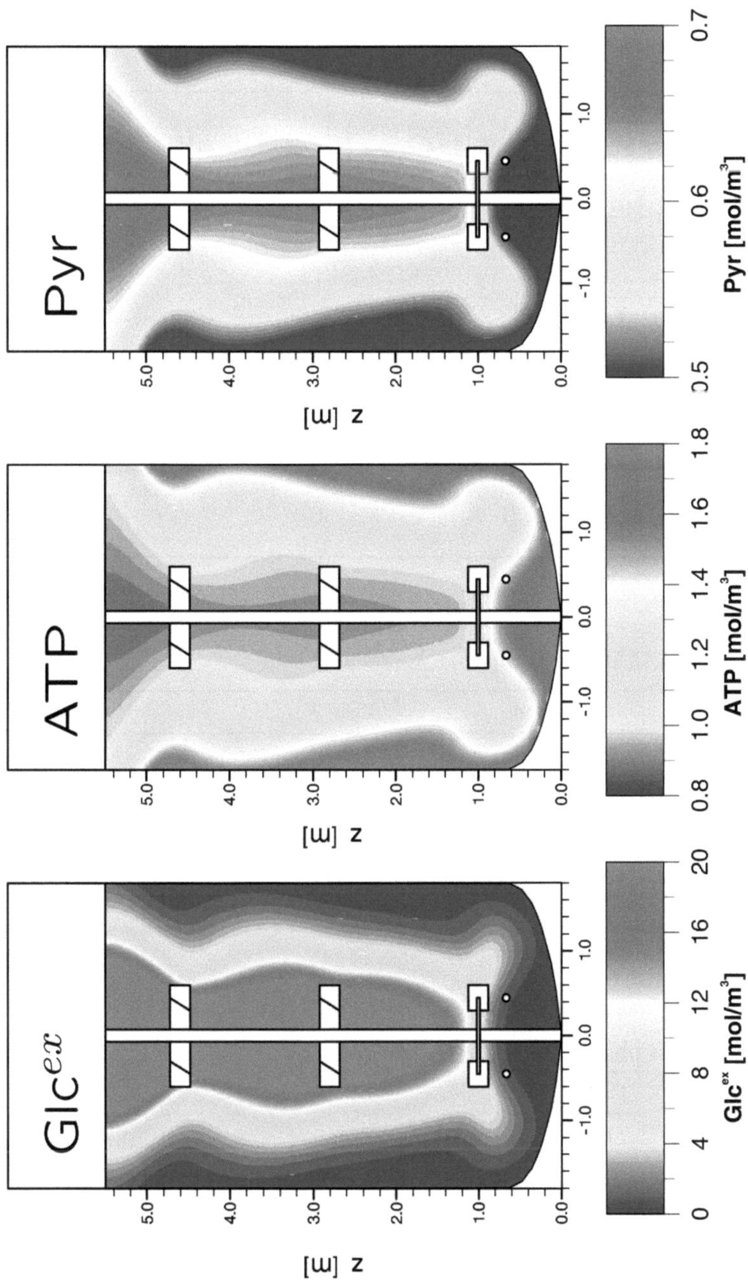

Fig. 27. Concentration distributions of glucose, ATP, and pyruvate four seconds after a glucose pulse onto the liquid surface

Table 5. Kinetic expressions in the model of *Bacillus subtilis* (Moes [78])

Rate	Kinetic expression
$r_{ATP \to X}$	$\left[\left(Y_{ATP,aer}\left(\frac{1}{Y_{P/S}}-\frac{1}{Y_{Ac/S}}\right)+Y_{ATP,anaer}\right)r_{S \to AC}+Y_{ATP/Bu}\left(r_{Ac \to Bu}-r_{Bu \to Ac}\right)R_S\right] \Big/ \left(1+\frac{Y_{ATP,aer}\,Y_{X/ATP}}{Y_{X/S}}\right)$
$r_{S \to E}$	$\left(\frac{1}{Y_{P/S}}-\frac{1}{Y_{Ac/s}}\right)r_{S \to Ac}-\frac{Y_{X/ATP}}{Y_{X/S}}r_{ATP \to X}$
r_{NADH}	$Y_{NAD,resp}\,r_{S \to E}+Y_{NAD/Ac}\,r_{S \to Ac}+r_{Bu \to Ac}-r_{Ac \to Bu}+r_{ATP \to X}\,Y_{X/ATP}\,Y_{NADH/X}$
$r_{S \to Ac}$	$k_1\,R_S$
$r_{Ac \to Bu}$	$k_2\,R_{Ac}\,R_S+k_{2,eq}\,R_{Ac}\left(1-R_S\right)$
$r_{Bu \to Ac}$	$k_3\,R_{Bu}\,R_S+k_{3,eq}\,R_{Bu}\left(1-R_S\right)$

$$R_S=\frac{c_S}{K_S+c_S} \qquad R_{Ac}=\frac{c_{Ac}}{K_{Ac}+c_{Ac}} \qquad R_{Bu}=\frac{c_{Bu}}{K_{Bu}+c_{Bu}} \qquad R_{O_2}=\frac{c_{O_2}}{K_{O_2}+c_{O_2}}$$

$$k_1=k_{1,0}+m_1\,R_{O_2} \qquad k_2=k_{2,0}+m_2\,R_{O_2} \qquad k_3=k_{3,0}+m_3\,R_{O_2}$$

Yield coefficient	Expression
$Y_{P/S}$	$Y_{P/S,0}+m_4 R_{O_2}$
$Y_{ATP/MADH_2}$	$Y_{ATP,0}+ATP_{\max}R_{O_2}$
$Y_{ATP,\,anaer}$	$Y_{ATP/PYR}+Y_{NAD,\,Ac}\,Y_{ATP/NADH_2}$
$Y_{ATP,\,aer}$	$Y_{ATP/PYR}+Y_{NAD,\,resp}\,Y_{ATP/NADH_2}$
$Y_{ATP,\,Bu}$	$Y_{NAD,\,Bu}\,Y_{ATP/NADH_2}$

oxygen in the liquid phase

$$S_{O_2^L}=-r_{NADH}\,Y_{O_2/NAD}\,c_X+k_L\,a\left(c_{O_2}^{*}-c_{O_2}\right) \tag{53}$$

oxygen in the gas phase

$$S_{O_2^G}=S_{O_2^L}\,\frac{1-\varepsilon_G}{\varepsilon_G} \tag{54}$$

Local mass transfer gas-liquid is again predicted with Eqs. (41)–(44). The kinetic expressions are summarized in Table 5. For further details the reader is advised to study the original papers [78, 79].

Figure 28 shows typical results from the simulation of the system of material balance equations which are parameterized with the results from the CFD simulations as described before. In this figure, a snapshot of distributions of dissolved oxygen concentrations and the production rate of butanediol after eight hours of simulated batch fermentation are shown. Figure 29 illustrates a comparison of measured [76] and simulated ratios of the two products acetoin

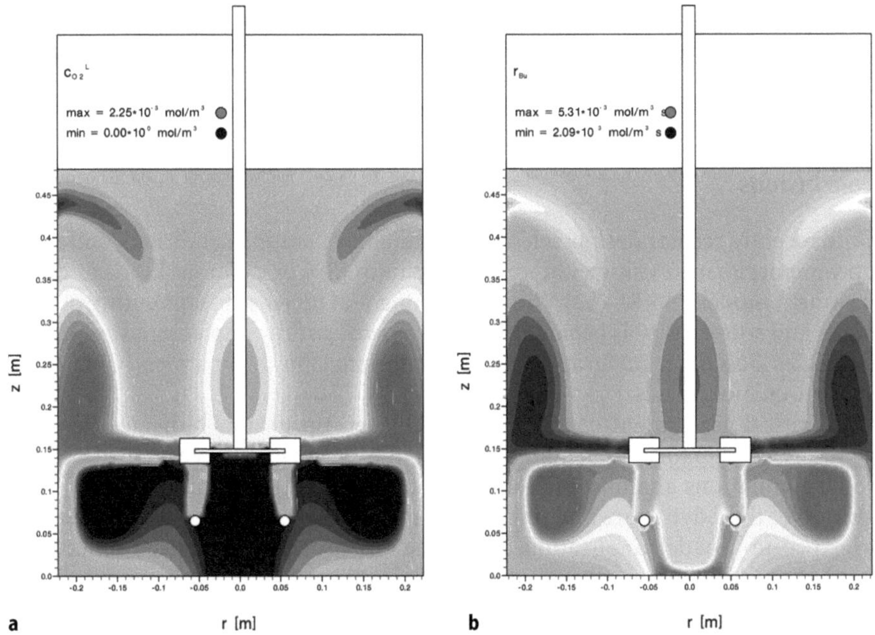

Fig. 28. **a** Dissolved oxygen concentration and **b** local production rate of butanediol at $t = 8$ h of a batch fermentation of *Bacillus subtilis*

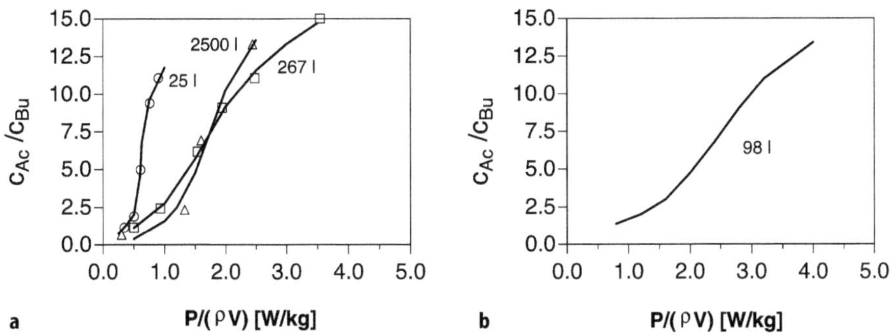

Fig. 29. **a** Measured [77] and **b** simulated final product ratios of acetoin/butanediol as a function of specific power input during fermentation

and butanediol at the end of the fermentation as a function of specific power input. The simulations show a behavior qualitatively similar to that observed by Griot [76].

6 Conclusions

Until recently, reactor design, selection of suitable operating conditions, and scale up were performed using either rules of thumb [80] or different kinds of compartment models [36, 81–83]. With the exponential increase in computing power, hard- and software tools became available to successfully implement simulation strategies based on integration of computational fluid dynamics (CFD) and structured biokinetics.

A critical assessment of the success achieved up to now indicates that the whole issue remains challenging. Despite the promising success of matching some simulations and observations and thus extending our knowledge in integration of fluid dynamics and physiology, more fundamental research is needed to expedite broader application of this new approach. Increased emphasis is needed on developing efficient and user-friendly software packages for direct simulations of fluid dynamics, incorporating phenomena of bubble coalescence and redispersion (population balances), and fundamental studies on integrating phenomena of micromixing in turbulent flow with biokinetics. The need for more detailed modeling of turbulence for this purpose has recently become a focus of fundamental fluid dynamics. Central among the many open questions is also a deeper understanding of the dynamics of the metabolic and regulatory networks as well as cascades of signal transduction triggered by external fluctuations.

Despite the many open problems to be tackled in the future, we expect the integrated approach to play an important role in advancing the performance of cell factories in technical bioprocesses.

7 References

1. Jenne M, Reuss M (1999) Chem Eng Sci 54:3921
2. Launder BE, Spalding DB (1972) Mathematical models of turbulence. Academic Press
3. Launder BE, Spalding DB (1974) Comp Meth Appl Mech Eng 3:269
4. Kim JJ (1978) Three dimensional turbulent flow-field in a turbine stirred tank. PhD thesis, Louisiana State University
5. Pope SB (1978) AIAA J 16:279
6. Hanjalick, Launder BE (1980) Trans ASME 102:34
7. Kline SJ, Cantwell BJ, Lilley GM (1981) The 1980–1981 HFOSR-HTMM-Stanford Conference on Complex Turbulent Flow, Stanford University, I, II, III
8. Roback R, Johnson BV (1983) NASA CR-168252
9. Chen YS, Kim SW (1987) NASA CR-179204
10. Patankar SV (1980) Numerical heat transfer and fluid flow. Mc Graw-Hill
11. Van't Riet K, Smith JM (1975) Chem Eng Sci 30:1093
12. Perng CY, Murthy JY (1993) AIChE Symp Ser 89:37
13. Takeda H, Narasaki K, Kitajima H, Sudoh S, Onofusa M, Iguchi S (1993) Comput Fluids 22:223

14. Brucato A, Ciofalo M, Grisafi F, Micale G (1998) Chem Eng Sci 53:3653
15. Placek J, Tavlarides LL (1985) AIChE J 31:1113
16. Gosman AD (1998) Trans I Chem Eng 76:153
17. Issa RI, Gosman AD (1981) The computation of three-dimensional turbulent two phase flows in mixer vessels. In: Numerical methods in laminar and turbulent flow. Pineridge Press, Swansea, p 829
18. Trägardh C (1988) A hydrodynamic model for the simulation of an aerated agitated fed-batch fermentation. In: Bioreactor fluid dynamics. Elsevier, p 117
19. Politis S, Issa RI, Gosman AD, Lekakon C, Looney MK (1992) AIChE J 38:1946
20. Morud K, Hjertager BH (1993) Computational fluid dynamics simulations of bioreactors. In: Mortensen U, Noorman H (eds) Bioreactor performance. IDEON, Lund, Sweden, p 47
21. Ishii M (1975) Thermo-fluid dynamic theory of two-phase-flow. Eyrolles
22. Ranade VV, van den Akker HEA (1994) Chem Eng Sci 49:5175
23. Kuo JT, Wallis GB (1988) Int J Multiphase Flow 14:547
24. Ishii M, Zuber N (1979) AIChE J 25:843
25. Drew DA, Lahey TJ (1987) Int J Multiphase Flow 13:113
26. Lopez de Bertodano M, Lahey TJ, Jones OCC (1994) Trans SME 116:128
27. Watanabe T, Hirano M, Tanabe F, Kamo H (1990) Nuclear Engng Design 120:181
28. Huang B (1989) Modelisation numérique d'écoulements disphasiques à bulles dans des réacteurs chimiques. PhD thesis, Lyon
29. Kowe R, Hunt JCR, Hunt A, Couet B, Bradbury LJS (1988) Int J Multiphase Flow 14:587
30. Lopez de Bertodano M. Lee SJ, Lahey RT, Drew DA (1990) ASME J Fluids Enging 112:107
31. Svendsen HF, Jakobsen HA, Torvik R (1992) Chem Eng Sci 47:3297
32. Johansen ST, Boysan F (1988) Metall Trans B 19B:755
33. Lahey RT, Lopez de Bertodano M, Jones OC (1993) Nuclear Enging Des 141:177
34. SatoY, Adatomi M, Sekoguchi K (1981) Int J Multiphase Flow 7:167
35. Rousar I, van den Akker HEA (1994) Proceedings of the 8th European conference on mixing, Cambridge, UK, p 89
36. Reuss M, Bajpai R (1991) Stirred tank models. In: Schügerl K (ed) Biotechnology, a multivolume comprehensive treatise, vol 4, measuring, modelling and control. VCH, Weinheim, p 299
37. Joshi JB, Pandit AB, Sharma MM (1982) Chem Eng Sci 37:813
38. Bombac A (1994) PhD thesis, University of Ljubljana
39. Bombac A, Zun I, Filipic B, Zumer M (1997) AIChE J 43:2921
40. Hinze JO (1955) AIChE J 3:289
41. Bakker A, van den Akker HEA (1994) Trans I Chem Eng 72:594
42. Greaves M, Barigou M (1988) Proceedings of the 6th European conference on mixing, Pavia, Italia, p 313
43. Jenne M (1999) Modellierung und Simulation der Strömungsverhältnisses in begasten Rührkesselreaktoren. PhD thesis, Universität Stuttgart
44. Reuss M, Schmalzriedt S, Jenne M (2000) In: Schügerl, Bellgardt (eds) Bioreaction engineering. Springer-Verlag (in press)
45. Bouafi M, Roustan M, Djebbar R (1997) Mixing IX, multiphase systems. Récents Progrès en génie des procédés 11(52):137
46. John AH, Bjalski W, Nienow AW (1997) Mixing IX, multiphase Systems. Récents Progrès en génie des procédés 11(52):169
47. Nienow AW, Elson TP (1988) Chem Eng Res Des 66:5
48. Cooke M, Middleton JC, Bush JR (198) Proceedings of the 2nd international conference on bioreactor fluid dynamics, BHRA/Elsevier, p 37
49. Abradi V, Rovera G, Baldi G, Sicardi S, Conti R (1990) Trans I Chem E 68:516
50. Harvey PS, Greaves M (1982) Trans I Chem Eng 60:201
51. Friberg PC (1988) PhD Thesis, NTNU Trondheim, Norway
52. Noorman H (1993) Bioreactor performance on 30 m^3 scale: verification of a scale-down/CFS approach. Technical report, Instituttet for Bioteknologi, Damarks Tekniske Hojskole, Lyngy, Denmark

53. Cui YQ, van der Lans RGJM, Noorman HJ, Luyben k ChAM (1996) Trans/Chem E 74:261
54. Voncken RM (1966) Circumlatie stromingen en menjing in geroerde vaten. PhD thesis, Delft University of Technology
55. Hoogendoorn CJ, Hartog AP (1967) Chem Eng Sci 22:1689
56. Landau J, Prochazka J (1961) Coll Czechoslov Chem Commun 26:1976
57. Khang SJ, Levenspiel O (1976) Chem Eng Sci 31:569
58. Tatterson GB (1991) Fluid mixing and gas dispersion in agitated tanks. McGraw Hill, New York
59. Groen DJ (1994) Macromixing in bioreactors. PhD thesis, Delft University of Technology
60. Bajpai R, Reuss M (1982) Can J Chem Eng 60:384
61. Kawase Y, Moo-Young M (1990) Chem Eng I 43:B19
62. Van't Riet K (1979) Ind Eng Chem Proc Des Dev 18:367
63. Theobald U, Mailinger W, Reuss M (1998) Anal Biochem 214:31
64. Rizzi M, Theobald U, Querfurth E, Rohrhirsch T, Baltes M, Reuss M (1996) Biotechnol Bioeng 49:316
65. Theobald U, Mailinger W, Baltes M, Rizzi M, Reuss M (1997) Biotechnol Bioeng 55:305
66. Vaseghi S, Baumeister A, Rizzi M, Reuss M (1999) Metabolic Eng 1:128
67. Schaefer U, Boos W, Takors R, Weuster-Botz D (1999) Anal Biochem 270:88
68. Buziol S, Bashir I, Baumeister A, Claasen W, Noisommit-Rizzi N, Mailinger W, Reuss M (2002) Biotechnol Bioeng 80:632
69. Mailinger W, Baumeister A, Reuss M, Rizzi M (1998) J Biotechnol 63:155
70. Rizzi M, Baltes M, Theobald U, Reuss M (1997) Biotechnol Bioeng 55:592
71. Mauch K, Hieber S E, Reuss M (2000) Proceedings of the 4th international congress on biochemical engineering, Stuttgart, Fraunhofer IRB Verlag, ISBN 3-8167-5570-4:57
72. de Koning W, van Dam K (1992) Anal Biochem 204:118
73. Liao JV, Hou S-Y, Chao Y-P (1996) Biotechnol Bioeng 52:129
74. Schäfer K, Boos W, Takors R, Weuster-Botz D (1999) Anal Biochem 270:88
75. Kaback H R (1969) Physiology 63:724
76. Larsson G, Törnkvist M, Stahl Wernersson E, Trägardh C, Noorman H, Enfors S-O (1996) Bioproc Eng 14:281
77. Griot M (1987) Maßstabsvergrößerung von Bioreaktoren mit einer sauerstoffempfindlichen Testkultur. PhD Thesis, ETH Zürich
78. Moes J (1985) Untersuchung von Mischphänomenen mit Hilfe von Bacillus subtilis. PhD Thesis, ETH Zürich
79. Moes J, Griot M, Keller J, Heinzle E, Dunn LJ, Bourne JR (1985) Biotechnol Bioeng 27:482
80. Kossen NWF (1992) In: Vardar-Sukan F, Suha Sukan S (eds) Recent advances in biotechnology. NATO Asi series, Kluwer Academic Publisher, p 147
81. Cui YQ, van der Lans RGJM, Noorman HJ, Luyben KCAM (1996) Trans IChemE 74(A):261
82. Alves S, Vasconcelos JMT, Barata J (1997) Trans IChemE 75(A):334
83. Vrabel P, van der Lans RGJM, Cui YQ, Luyben KCAM (1999) Trans IChemE 77(A4):291
84. Chassagnole C, Noisommit-Rizzi N, Schmid J-W, Mauch K, Reuss M (2002) Biotechnol Bioeng 79:53

Received: February 2002

A 'Fine' Chemical Industry for Life Science Products: Green Solutions to Chemical Challenges

A. Bruggink[1,2] · A. J. J. Straathof[3] · L. A. M. van der Wielen[3]

[1] DSM Research, P.O. Box 18, 6160 MD Geleen, the Netherlands
[2] University Nijmegen, Department of Organic Chemistry, Toernooiveld, 6525 ED Nijmegen, the Netherlands
[3] Kluyver Laboratory for Biotechnology, Delft University of Technology, Julianalaan 67, 2628 BC Delft, the Netherlands. *E-mail: L.A.M.vanderWielen@tnw.tudelft.nl*

Modern biotechnology, in combination with chemistry and process technology, is crucial for the development of new clean and cost effective manufacturing concepts for fine-chemical, food specialty and pharmaceutical products. The impact of biocatalysis on the fine-chemicals industry is presented, where reduction of process development time, the number of reaction steps and the amount of waste generated per kg of end product are the main targets. Integration of biosynthesis and organic chemistry is seen as a key development.

The advances in bioseparation technology need to keep pace with the rate of development of novel bio- or chemocatalytic process routes with revised demands on process technology. The need for novel integrated reactors is also presented. The necessary acceleration of process development and reduction of the time-to-market seem well possible, particularly by integrating high-speed experimental techniques and predictive modelling tools. This is crucial for the development of a more sustainable fine-chemicals industry.

The evolution of novel 'green' production routes for semi-synthetic antibiotics (SSAs) that are replacing existing chemical processes serves as a recent and relevant case study of this ongoing integration of disciplines. We will also show some challenges in this specific field.

Advances in Biochemical Engineering/
Biotechnology, Vol. 80
Series Editor: T. Scheper

1 Introduction

The fine-chemicals industry manufactures (ingredients for) life saving drugs, for healthy nutrition and for consumer products, that increase the overall well being of mankind. The annual sales of fine-chemical products are estimated to be at US$ 40 billion worldwide in 2000. This industry employs hundreds of thousands of workers, scientists and engineers. It is an important player in national and international economies, and it is expected to continue doing so in the future. Economic competitiveness, product quality control as well as care for the environment and natural resources provide important constraints and goals for the development of the fine-chemicals industry. Truly sustainable and feasible processes need to be developed in an integrated form. This process integration can occur at different levels: at a molecular, equipment or process scale. Integration can also proceed at company level through acquisition, which is often a start for process integration as is discussed here.

1.1 Molecular Integration

Many fine-chemical industries embrace biotechnology (biocatalysis, biotransformations and fermentation technology) in combination with catalytic organic synthesis to replace traditional stoichiometric processes to grow towards greater sustainability. Optimal solutions may require the *integration at the molecular level*, namely of the underlying of bio- and chemocatalysis processes. This requires the screening for new biocatalysts that are active under novel and often non-natural conditions. In some cases, simple reactors according to the "single pot" concept may be feasible. The conditions have greatly enhanced the successful introduction of biocatalysis in the fine-chemicals industry. For further growth, it is expected however, that novel conditions lead to novel demands on process technology.

1.2 Multifunctional or Integrated Equipment

When reactions are reversible or products unstable, it is attractive to integrate recovery and (bio-)reaction, that is in situ product removal (ISPR). Compatibility of bioconversion and separation conditions is a key issue in ISPR. It will be demonstrated in a later section that constraints in an integrated system are completely different from those in the individual, non-integrated process steps. It may also be attractive to combine separation steps. A well-known example is crystallization with a withdrawal of coarse crystals (integration of molecular and mechanical separations). Often, an optimal integrated system will operate under conditions that are not equal to those of the individual and non-integrated conversion and separation steps. This is process integration at the level of *unit operations*.

1.3 Integration at the Plant Level

Conversions are seldom complete and fully selective towards the target product. This requires high-resolution purification techniques. Many conventional technologies such as chromatography and crystallization may provide solutions; however, rational selection of separation steps and their order in a cascade, their fast development, and tuning also requires an integrated approach. The individual stages need to be optimised but also the overall integrated process, including the reaction steps. This is process integration at the level of the *complete plant*. To analyse complete processes, one has to balance capital costs of new investments versus variable costs of running plants (usually complex, costly equipment leads to a reduction of the variable costs), but also different sorts of auxiliary streams (materials and energy) have to be balanced.

The last of these requires tools such as exergy analysis in addition to process economics.

1.4 Process Integration Should Start in the R&D Laboratories

Many processes are still performed batchwise and frequently in a single stage. This is in many cases far from the technological and economic optimum. The basic reason is that many industrial fine-chemicals processes are scaled-up versions of the original laboratory equipment in which a batchwise and step-by-step approach is always the start of development. It is evident that this is a result of a constant pressure to reduce the time-to-market, confidence in proven technology, the prejudice that novel technology is always more expensive, and an incomplete set of technological tools for high-speed process development. It is also evident that this routine of process development needs to change for a number of reasons:

(1) Many established biotechnological and pharmaceutical products are losing patent protection. Therefore, price competition and cost efficiency, in manufacturing as well as in scientific R&D, will play an increased role in maintaining competitiveness. A major leap forward in process technology will enable renewed protection of second and higher generation processes. This has occurred for racemic pharmaceuticals, which after a "chiral switch" could again be protected, as the new processes produced enantiomerically pure pharmaceuticals.

(2) The environmental burden of small- and large-scale processes has to be reduced as much as possible. Waste reduction (mass and energy) of course has an ethical component, but also economic competitiveness dictates that cleaner solutions are found. Auxiliary materials, including their regeneration or disposal costs, may contribute significantly to the cost price of the product. An example is the production of recombinant insulin by *E. coli* fermentation as is described by Datar and Rosén [1]. The auxiliary materials in the downstream processing were estimated to contribute approximately 12% of the production costs, and waste treatment to approximately 5%.

(3) Batch processes are inherently dynamic and more difficult to monitor, control and optimise than steady state, continuous processes. Quality control of the product in a dynamic process is more difficult to achieve. Also, process safety is more difficult to achieve in a dynamic system than in continuous production.

(4) In fine-chemicals production, plants are often multipurpose for reasons of flexibility. ISPR is difficult to achieve in non-dedicated equipment, particularly when it is operated batchwise. For instance, reactive distillations can virtually only be achieved in a dedicated, steady state system. Control over the crystal quality (composition and particle size distribution) in a reactive crystalliser is practically impossible when the concentrations of product, substrates and contaminants vary widely.

It is a tremendous challenge for process engineers, as well as chemists and life scientists to generate "green" integrated technological solutions for the fine-chemicals industry. For a sustainable fine-chemicals industry, however, in a modern, developed world, all issues mentioned above need to be addressed, preferably simultaneously. In this contribution, we will discuss process integration aspects at these fairly different levels. We will also illustrate the possibilities and their impact on manufacturing processes for various penicillins and cephalosporins.

2 Discussion

2.1 Conversion Technology

The introduction of biocatalysis in the synthesis of industrial chemicals, in particular fine-chemicals, can be seen as a first step in the integration of organic synthesis and biosynthesis. Nowadays, a large number of biocatalysts are being applied in industry and an overview of the specific types is given in Fig. 1. The onset of this development is due to the need to replace traditional stoichiometric processes by catalytic processes with improved product-to-waste ratios [2]. The cumbersome translation of petrochemical catalysis to catalysis for the more complex fine-chemical molecules has favoured the fast acceptance of biocatalysis and biotransformations.

Although (asymmetric) chemical catalysis allowing reactions at ambient temperatures is developing fast, biocatalysis is in the lead from an industrial point of view. A development from single and relatively simple enzyme-catalysed conversions to more complex biotransformations, employing a number of enzymes, including cofactors, effecting multistep "single-pot" processes is well underway. As is shown in Fig. 2, the integration of organic synthesis and fermentations might be the end result, indeed a green chemistry.

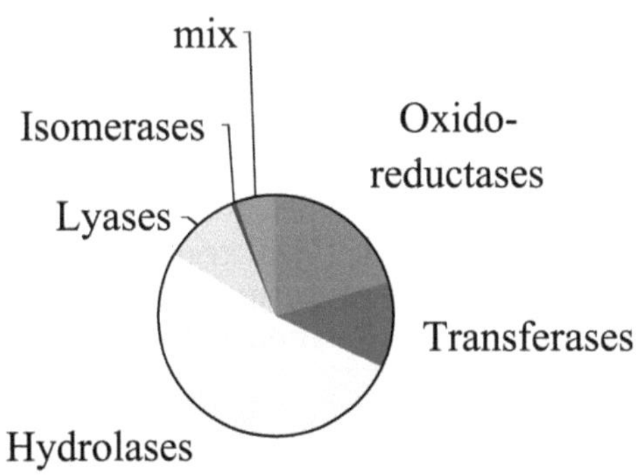

Fig. 1. Overview of the types of enzymes used in around 100 commercialised biotransformations [3]

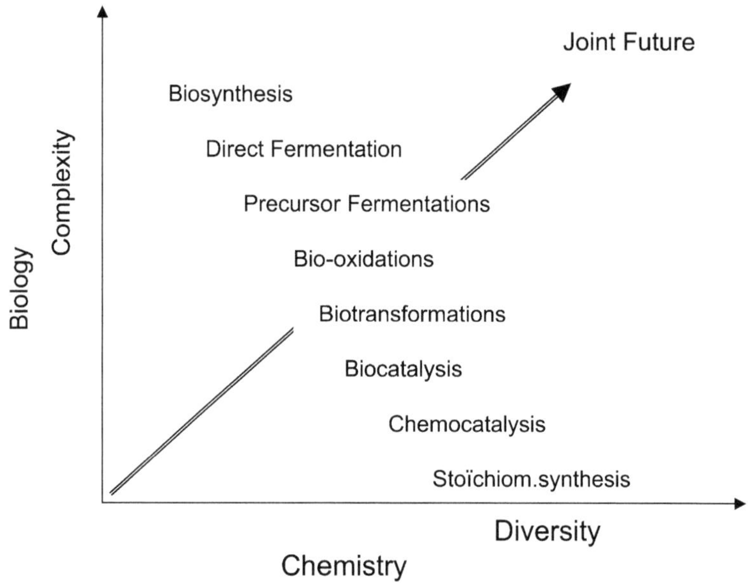

Fig. 2. Synthesis from chemical and biological perspective (after J. M. Lehn)

2.1.1
Hydrolysis and Synthesis

An analysis of the commercialised biotransformations (Fig. 1) shows that in 50% of the processes only hydrolases are being used. This is in line with the analysis of Faber [4], who showed that about 60% of the research publications on biocatalysis deal with hydrolases. The reason for this is partly that making a molecule is more difficult than breaking a molecule. However, hydrolases are also used in the reverse mode, pulling the equilibrium towards synthesis by water removal during the reaction, for example

$$\text{amine} + \text{carboxylic acid} \rightarrow \text{amide} + \text{water} \qquad \text{(a)}$$

Clearly, such a thermodynamically controlled reaction should preferably be performed in the absence of water. Therefore, the study of the stability and activity of enzymes under non-aqueous conditions remains a key issue in biocatalysis. The systematic study of reaction and phase equilibria is important as well, because it may lead to the identification of reactions that previously were assumed to be thermodynamically not feasible [5], or reaction conditions that were assumed to be not feasible [6]. In these cases, suspended substrates or products are used. In a later section, product precipitation will be treated from the viewpoint of in situ product removal.

The large flexibility that hydrolases show towards conversion of unnatural substrates is an advantage when compared to other types of enzymes. For instance, instead of water, hydrolases can use ammonia, amines, alcohols and many other nucleophiles. This allows them to be used as "transferases". If the aforementioned amide synthesis by reverse hydrolysis is thermodynamically unfavourable under the conditions where the enzyme activity is good, the amide may be synthesised in better yield by the same hydrolase by using an activated substrate, such as the methyl ester of the carboxylic acid:

$$\text{amine} + \text{methyl carboxylate} \rightarrow \text{amide} + \text{methanol} \quad \text{(b)}$$

In contrast to the true transferases, competition between water and the non-natural nucleophile for reaction with the enzyme-acyl species will occur, leading to undesired loss of the activated substrate and of the product:

$$\text{water} + \text{methyl carboxylate} \rightarrow \text{carboxylic acid} + \text{methanol} \quad \text{(c)}$$

$$\text{water} + \text{amide} \rightarrow \text{carboxylic acid} + \text{amine} \quad \text{(d)}$$

Because of these undesired reactions, the maximum yield of amide is not reached at thermodynamic equilibrium but at an intermediate stage. As this maximum yield is determined by the enzyme kinetics, the reaction is said to be kinetically controlled.

The choice of a thermodynamically or kinetically controlled synthesis not only has important implications for the study of the reaction conditions, but also on the development of the enzyme, the reactor and even on the downstream processing, as ISPR (in situ product removal) will be important (see Table 1).

The table clearly shows that thermodynamically controlled reactions are inherently simpler. Their only drawback is that at the thermodynamically most favourable conditions, there may be severe kinetic limitations and the reaction will be too slow. These kinetic limitations may be partly due to mass transfer. For example, the dissolution rate of a solid substrate may be too low in a non-aqueous medium. However, to a large extent the kinetic limitations will be caused by

Table 1. Comparison of thermodynamically and kinetically controlled enzymatic synthesis reactions

	Thermodynamic control	Kinetic control
Substrate characteristics	Cheap	Activated substrate required
Reaction condition optimisation	Use thermodynamic data	Use kinetic data
Reactor optimisation	Different reactors give similar yield	Reactors with least back mixing give highest yield
Monitoring of reaction	Not important; yield will increase to maximum	Important; yield will go through maximum
Enzyme development	Active enzyme needed at sometimes unfavourable conditions	Active enzyme needed; continuous drive to develop more selective enzyme
Enzyme immobilisation	Has little influence	Diffusion limitation may decrease selectivity
ISPR target	Water or product removal	Product removal

absence of appreciable enzyme activity under non-natural conditions. As modern screening and protein engineering techniques seem to lead to a supply of enzymes that are suited for non-natural conditions, the long-term targets for screening and protein engineering should be based on enzymatic activity at conditions set by the thermodynamically controlled reaction.

In kinetically controlled reactions, the main strategy is to reduce the amount of side-reaction with water. When this strategy is very successful, either by using non-aqueous conditions or by improving the selectivity of the enzyme up to the level that water is not recognized as a substrate anymore, one ends up at a situation that there is only a single, thermodynamically controlled, reaction. Thus, a kinetically controlled reaction, when improved continuously, ultimately could become a thermodynamically controlled reaction.

2.1.2
Redox Reactions

In the chemical industry, oxidation and reduction reactions are preferably carried out with cheap electron acceptors are donors, such as O_2 and H_2, respectively. If a complicated molecule is to be oxidized or reduced, different products may be obtained, depending on the selectivity of the catalyst. For the synthesis of fine-chemicals, the selectivity of chemocatalysts is not always sufficient and biocatalysts provide a very interesting alternative. However, relatively few redox enzymes use O_2 or H_2 as one of the substrates; these few enzymes are popular targets as catalysts for fine-chemicals production. However, the electrons in enzymatic redox reactions are usually accepted or provided by a coenzyme, which is most often the oxidized or reduced form of NAD(P). The development of processes involving the efficient regeneration of the converted coenzyme has been subject of much research. Two types of biological redox processes are being applied, either using isolated enzymes or using microorganisms.

Isolated enzymes are used mainly for reductions, using regeneration of NADH by formate dehydrogenase (FDH) [7]. The advantage of this regeneration reaction is that formate is a relatively cheap electron donor, and the overall reaction is driven to completion because carbon dioxide is liberated. For reduction of a ketone to a (chiral) alcohol using NAD-dependent ADH (alcohol dehydrogenase), the simultaneous reactions are:

$$\text{ADH-catalysed:} \quad \text{ketone} + \text{NADH} + H^+ \Leftrightarrow \text{alcohol} + \text{NAD}^+ \tag{e}$$

$$\text{FDH-catalysed:} \quad \text{NAD}^+ + \text{formate} \rightarrow H^+ + CO_2 + \text{NADH} \tag{f}$$

FDH from *Candida boidinii* is being produced at pilot scale and is available in significant quantities. Therefore, this reaction can be generally used for NADH regeneration. Recently, the same concept has been used for NADPH regeneration. An NADPH-dependent FDH has been obtained by multipoint site-directed mutagenesis of the gene coding the enzyme from the bacterium *Pseudomonas* sp. 101 [8]. For efficient shuttling of the redox cofactor between the two enzymes, proper reaction conditions have to be maintained. These are most easily maintained in a continuous stirred tank system, in which the enzymes and coenzymes are retained using an ultrafiltration membrane [7]. This membrane also takes away the need to remove pyrogens in the downstream processing.

Instead of using combinations of enzymes and coenzymes, it should be possible to have a single enzyme that performs the overall reaction:

$$\text{ketone} + \text{formate} \rightarrow \text{alcohol} + CO_2 \quad \text{(g)}$$

Recently, it has been shown that there are NAD-dependent oxidoreductases that will not liberate NADH/NAD^+ from the active site. They catalyse such redox reactions, albeit not with formate but with less favourable electon donors and with low rates [9]. When these enzymes can be properly engineered and produced, they will impose few constraints on the reactor design. This situation is analogous to what has been described in the previous section for synthetic reactions using hydrolases: if a biocatalyst is found that can directly convert the substrates into the desired products, without formation of intermediates or occurrence of side-reactions, the reactor design becomes simple.

Although a coenzyme-regeneration system using FDH is feasible, it may not always be worthwhile to find, produce and purify the required enzymes, and to build a dedicated reactor. Instead, regeneration is often carried out with living cells, requiring only one fermentation to obtain the desired biocatalyst. Then, regeneration can be carried out with a cheap substrate and the enzymes present in the whole cells, such as alcohol dehydrogenase, in the following reaction:

$$NAD^+ + \text{ethanol} \rightarrow H^+ + \text{acetaldehyde} + \text{NADH} \quad \text{(h)}$$

Usually a large number of other reactions will occur simultaneously, some of them being beneficial for the coenzyme regeneration, whereas others lead to undesired by-products. Also, the substrate and product of the main reaction may get involved in undesirable side-reactions. Therefore, whole-cell reactions may be cheaper and simpler to carry out than reactions using isolated enzymes, but they are less easily controlled, less reproducible and yield more waste. A well-known example of this type is the reduction of ketone derivatives catalysed by *S. cerevisiae* (baker's yeast). This microorganism is very cheap and generally available [10]. Due to the elucidation of the genome of baker's yeast it is becoming very attractive to knock out undesired enzyme activities and to amplify the desired activities [11]. However, the outcome of such an approach can be surprising, as the physiology of microorganisms is far from being completely understood. Metabolic engineering approaches that try to elucidate the complete cell energetics will be required to progress in the area of whole-cell biocatalysis.

At the same time, engineering rules that apply to whole-cell redox reactions have to be taken into account. In general, aeration and/or carbon dioxide production is involved, and plug flow reactors are not appropriate. Moreover, oxygen transfer to immobilized cells is not very efficient. Consequently, continuous reactors are not very suitable for redox biotransformations with whole cells [12]. These biotransformations can best be carried out in (fed) batch reactors with free cells. Since the production of the cells will also involve a (fed) batch process, these processes may easily be combined. Then, the cells are produced in a fed-batch fermentor (growth stage), and if the biomass concentration has reached a sufficiently high level, the precursor of the biotransformation may be added and its oxidation or reduction will be started (biotransformation stage).

2.1.3
Lyases and Transferases

For catalysing thermodynamically controlled reactions, lyases and transferases provide clear opportunities. Their relatively narrow substrate specificity largely prevents the occurrence of side-reactions, although at the same time this limits their applicability to compounds that are fairly closely related to their natural products. However, in some cases these products are synthetically very valuable, for example when carbon-carbon bonds are formed in an enantioselective manner.

A reasonable number of biotransformation processes using lyases or transferases have been developed on an industrial scale, but this has not yet led to a general picture about the best process configuration. The main problems that seem to occur with these reactions (unfavourable equilibria and instability of substrates or products) have been solved in different manners. Substrates are fed slowly into the reactor or dissolved gradually, products are removed in situ by extraction or crystallization, or the biotransformation enzyme is incorporated in a cascade of reactions using whole cells. Thus, either of the aforementioned approaches seems to be feasible, given a specific biotransformation.

2.1.4
Development of Novel Biocatalysts

For all important types of biotransformations, it can be expected that soon there will be rules of thumb that allow the rapid selection of the preferred reactor type, using some basic characteristics of the reactants and biocatalyst only. In such a situation there is a limited need to optimise the reaction conditions by using a mechanistic model, as the main value of a mechanistic model is its power to predict the effect of an extrapolation. When the reaction type is fixed, only prediction of the effect of an interpolation is required, and this can be done with a black-box model using a data-driven analysis. Due to the availability of useful algorithms for experimental design and optimisation, process development may be speeded up considerably in this manner. When such a situation is reached, there would be a clear resemblance to the development of protein engineering. Originally, this was mainly performed by rational optimisation, but presently random techniques are preferred because they have a higher success rate. The development of robotized screening methods and powerful optimisation algorithms is a key factor in the success of random methods.

So far, industrially applied biocatalysts mainly serve hydrolytic reactions (Fig. 1). Later in this work, some industrial examples towards use in synthesis are also given. Although there are around 350 industrially available enzymes, this is still rather limited compared to the vast natural diversity.

Several research lines can potentially contribute to the advancement of biocatalysis and biotransformations in industrial applications:

- high-speed screening techniques for multitudes of natural or genetically engineered enzymes;
- reliable and speedy methods for the controlled industrial production of tailor made enzymes;
- fast and reliable methods to determine the structure of whole enzymes and in particular their active sites as well as the catalytic mechanism;
- rational formulation methods, that is immobilization, of enzymes to stable and robust industrial biocatalysts.

Fine tuning of enzyme formulations might increase the present number of industrially available enzymes from 350 to a few thousands biocatalysts. In particular, the development of new formulations that enhance selectivity, efficiency and stability is crucial. In addition, a closer collaboration between organic chemists and molecular biologists can lead to novel bio-inspired catalyst systems that combine the best of two worlds.

2.2
Separation Technology

Many fine-chemical products are intermediates or final products for the pharmaceutical industry. Therefore, demands on product purity and control of product purity are high. In particular, the levels of near-identical contaminants such as (stereo) isomers, degradation and by-products of the synthesis pathway, such as oxidation, cyclization and ring-opening products should be small. Contaminants with very similar molecular structures as the main product, may cause serious adverse responses when included in the final products.

Also for food ingredients, demands are becoming stricter in terms of purity and control of composition. This conflicts with a 'natural' image and minimal processing. Also the selection, contact and residual levels of auxiliary materials (solvents, salts, sorbents, etc) are restricted in this sense. Legislatory demands for these food ingredient products will also tighten, in particular for the novel Life Science Products, such as nutraceuticals. This will generate new demands and constraints for selectivity and efficiency of purification processes.

2.2.1
Some Basic Separation Theory

Single-stage, batch or continuous separation steps in multiphase systems can only lead to near-complete separations when the partition (or distribution) coefficients of the components over the various phases are sufficiently different. Because of the common structural similarity of main products and contaminants, this is usually not the case. The key parameter is the so-called separation factor S (Fig. 3). Assuming thermodynamic equilibrium between outlet flows of a single equilibrium stage, the separation factor S relates performance to the ratio of auxiliary flow (V) and feed flow (L) and the distribution coefficient of the

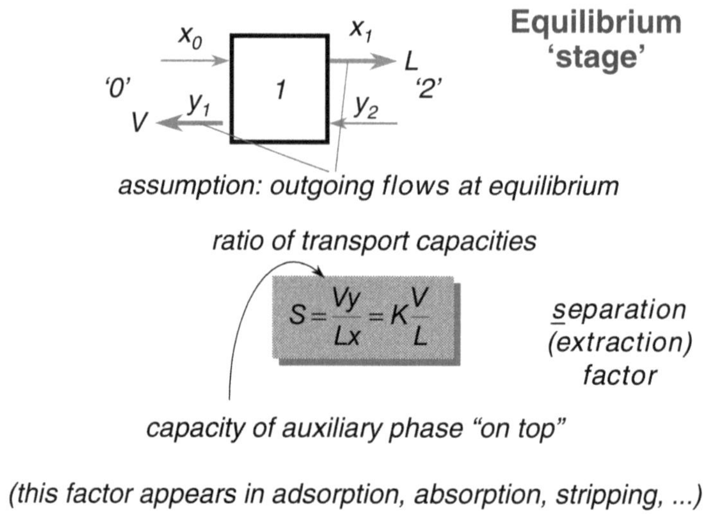

Fig. 3. Separation factor in a single equilibrium stage

component of interest (K, all in consistent units). The performance is measured in terms of the achieved change in concentrations x and y in either phase. When the amount of auxiliary phase or the partition coefficient increases, the separation factor increases and the degree of recovery in auxiliary flow or phase V increases as well. For single and multistage contact with constant partition coefficients, simple relations can be derived [13]. Multicomponent systems with more complex thermodynamics require rigorous models with numerical solutions for an adequate description. Calculations show that multistage, counter-current cascades with feed streams at either end of the cascade can improve the recovery largely. However, the selectivity of a separation can be improved only to a limited degree.

The separation factor S, also known as the extraction factor, is a measure for the ratio of carrying capacities of the flows for a specific solute. When $S > 1$, most of a species is transported with flow V; when $S < 1$, most of the species remains in flow L. This offers opportunities in the form of fractionating technology, to improve the performance to well above what can be achieved in single stage and (single section) multistage counter-current systems.

Kinetic separations, in which components are separated on the basis of different diffusive or convective velocities (Fig. 4) can lead to much higher resolutions in a single stage. These are shown as differently sized arrows in Fig. 4 to illustrate respectively membrane based and chromatographic separations. Unfortunately, many membrane separations are not yet sufficiently selective to discriminate between very similar molecules, such as protein mixtures. Although fixed bed chromatography is well suited to separate mixtures of similar components, often substantial flows are required to obtain sufficiently different convective velocities which leads to a substantial eluent consumption.

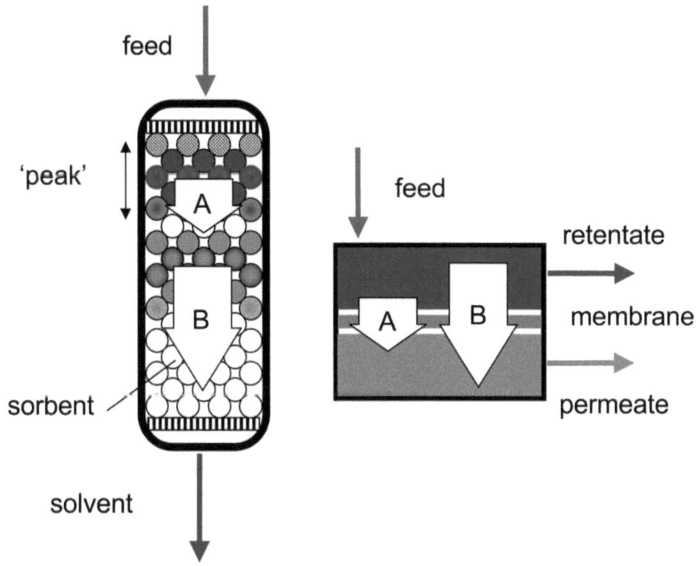

Fig. 4. Kinetic separations on the basis of differences in convective (chromatography) or diffusive (membranes) velocities

2.2.2
Fractionation Technology

Ideally, the desired product must be concentrated, while contaminants are simultaneously, spatially removed in a fractionating (chromatographic) manner. This concept of fractionating separations is most easily visualised for the binary separation of two similar components A and B with slightly different partition coefficients K_A and K_B ($K_A > K_B$) in any suited biphasic system. The flow rates of the two counter-current auxiliary phases are such that, component A moves primarily in the flow direction of flow V, whereas B moves primarily in the opposite flow direction, that is the flow direction of flow L. Flow L can be an aqueous stream composed of the (aqueous) feed optionally diluted with extra process water. V is a second phase or flow of the product itself (crystals, water immiscible liquid product), or an auxiliary stream of adsorbents, ion exchange resins and solvents. V may also be an aqueous stream separated from L by a membrane. Fractionation technology separates components introduced as a mixture at feed location F in Fig. 5, into two fractions at high yields, even when the partition coefficients are very similar. The basic configuration comprises two sections as is shown in Fig. 5.

Adequate, cost efficient and optimal operation can be achieved by reducing process streams and optimising concentrations. This may require more complex configurations with additional counter-current sections and reflux streams. A well-known classical form is the distillation column in which part of the top vapour and bottom liquid products are recycled (refluxed) to the column. This

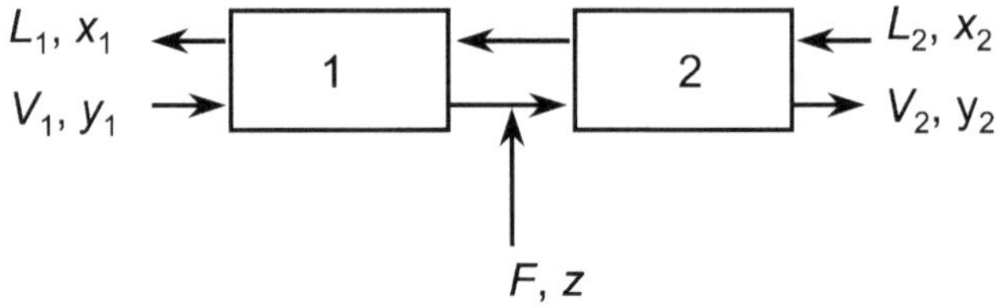

Fig. 5. The basic fractionating configuration, where two products introduced in stream F can leave the fractionator separated in streams L_1 and V_2. x and y indicate the compositions of the respective flows

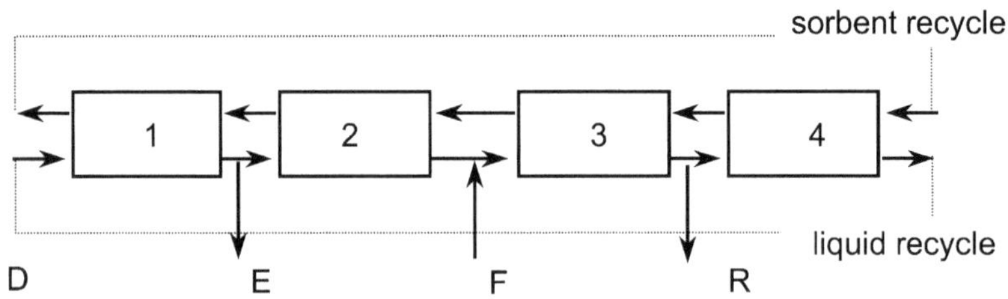

Fig. 6. Schematic diagram of an SMB system with four counter-current sections for the chromatographic fractionation of a mixture introduced in the feed stream F, into an extract product (E) and a raffinate product (R), using a desorbent stream (D)

enhances the purity of these products. An upcoming technology for the field of fine-chemicals production is the simulated moving bed technology (Fig. 6). The auxiliary flow V is an adsorbent flow, whereas the other phase is a fluid, usually a liquid (L). The basic configuration comprises two central sections (2 and 3), responsible for the actual fractionation, and two end sections (1 and 4) that regenerate the eluent (or desorbent, 1) and liquid (4) flows and increase the solute concentrations. The mixture of components in feed stream F is split into an extract fraction, leaving in flow E, and a raffinate fraction, leaving in flow R.

These fairly complicated systems can, under a number of simplifying assumptions, be described with a relatively small number of equations [13]. The common short-cut design procedure for the aforementioned fractionating systems of Figs. 5 and 6, for constant partition coefficients, follows the scheme outlined in Table 2.

We will not elaborate on the details of the design for different systems, but focus on the possibilities to perform difficult separations while minimizing the amount of auxiliary materials required. This concerns essentially step 2 of the design procedure from Table 2 and can be restricted to the basic fractionating configuration in Fig. 5. In the Appendix, the procedure is outlined in more detail. Here, we restrict ourselves to the outcome of the procedure in terms of an operating window for flow rate ratios m of liquid and sorbent streams $m_j = L_j/V_j$. In Fig. 7, such an operating window of flow rate ratios for a prefeed (m_1) and a postfeed (m_2) section in a fractionating unit is shown. The relevant window of oper-

Table 2. Short-cut design procedure for separation equipment

1. Identify relevant thermodynamic properties, such as distribution coefficients.
2. Select the flow rate ratios V and L in each of the sections, assuming the separation factors for each component in each of the sections to be smaller or larger than unity, according to the preferred direction of the component ($S > 1$: with V-flow; $S < 1$, with L-flow).
3. Determine hydraulic constraints which are given by maximum pressure drops in packed beds, by hindered rise or settling velocities in liquid-liquid or solid-liquid systems or by pressure balances in gas-liquid contactors. This leads in essence to the cross-sectional area of the contactor.
4. Calculate the required degree of contact between the two phases to allow sufficient mass transfer. This determines in essence the volume and length of the contactor. The Appendix shows underlying mathematical models and their general solution procedure for non-reactive and reactive systems.

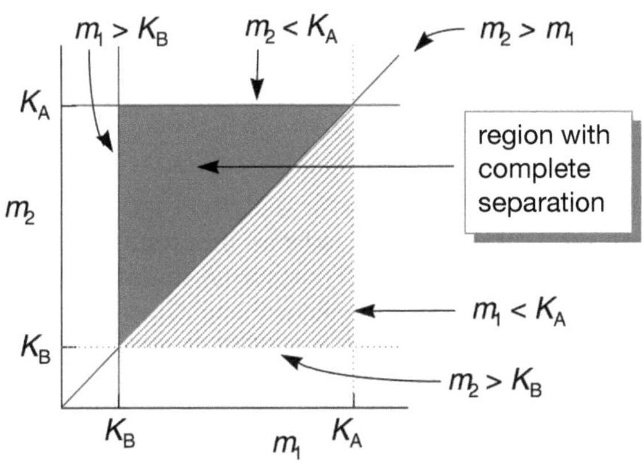

Fig. 7. Operating conditions for complete recovery of A (*'heavy key'*) and B (*'light key'*) products by a two-section fractionating separation

ating conditions relate to the shaded, triangular upper-diagonal area in the $m_1 - m_2$ plane.

Any point in the triangle can result in complete separation of a mixture of components A and B, as well as in complete recovery of each individual component (for instance A in the V-stream and B and the L-stream), provided that the sections contain sufficient numbers of the equilibrium stages. The optimal point for efficient usage of auxiliary material in the V-stream is represented by the upper left corner of the triangle. In this point, the difference between the flows of feed phase, which is the feed flow rate ($m_2 - m_1 = F/V$) is largest. Robust operation is effected by allowing a larger flow rate of V, in accordance with expected fluctuations in process operation (for instance by operating at a 10–30% higher consumption).

In Figs. 5 and 7, we specified neither the nature of the two counter-current phases (GL/LL/SL/SG/sorbent-L/sorbent-G etc.), nor the nature of the equipment used for contact. Hence, the methodology is fairly general and can in principle be applied to extraction, crystallisation, distillation, gas and liquid chromatography as well as to membrane separations.

Our simplified analysis indicates that a near-complete separation is possible, even for very similar components ($K_A \rightarrow K_B$). In the latter case, the price to pay is that the flows of the auxiliary (V) and "eluent" (L) phases may become excessive. Before relating actual flows to specific separation problems, we can estimate the minimal flows qualitatively. For instance, to recover products at concentrations in the product streams in four-section SMB-systems (Fig. 6), similar to their original (feed) concentrations, the "eluent" or "desorbent" flow should equal the feed stream [1]. Poorly soluble components, low capacities of the phases and near-identical partition coefficients, lead to large internal process streams and thereby to voluminous equipment and a substantial energy consumption.

Fractionating technologies are now upcoming for many biotechnological separation systems. The best-developed methodologies are continuous (resin-liquid) chromatography and extraction [14–16], and to a smaller extent, fractional crystallisation and membrane-aided separations [17, 18].

2.2.3
Chromatography

Chromatography is often associated with analytical and small-scale preparative separations, and is too often assumed to be an inherently batchwise and discontinuous fractionation technique. The continuous simulated moving bed (SMB) technology is the more efficient answer to these disadvantages of batch chromatography. The successful four-section SORBEX concept for SMB chromatography was originally developed for large-scale separations such as that of xylene isomers (system capacities up to 400 kton year^{-1}) and for sugar separations (system capacities up to 100 kton year^{-1}). It is currently increasingly implemented in the form of optimised, smaller systems suited for septic operation in the fine-chemicals, biotechnology and food specialty industries with modest production volumes. These systems allow operation at high pressure for HPLC and supercritical chromatography applications [19].

SMB technology now seems to be a well-accepted option for the separation of enantiomers. Whereas most systems are run in an isocratic manner (identical solvent composition of feed and eluent streams), novel operating procedures such as pressure gradient [19], solvent gradient SMB [20, 21] and salt gradient SMB [22] have shown new and general routes for the optimisation of these systems by minimisation of eluent and resin volumes several-fold. In Gradient SMB, the feed and the desorbent streams have a different solvent composition. The desorbent is richer in the better solvent, which lowers the partition coefficient of the stronger adsorbing species in the bottom sections. This facilitates desorption.

[1] This is true for dilute products with non-interacting linear isotherms. It is more accurate, especially for more concentrated products, to balance the solvent fractions in the feed and in the eluent streams.

Jensen et al. [21] and Houwing et al. [22] demonstrated a several-fold reduction in eluent consumption and resin inventory as well as a concentrating effect on the product in the extract flow.

Particularly useful are the so-called carousel type SMB systems. These systems may comprise more than four sections. The columns can be configured in parallel as well as in series. This increased flexibility allows for internal recycles as well as multiple feed and product flows. In this manner, multiple chromatographic actions are combined within a single piece of equipment.

2.2.4 Crystallisation

Most fine-chemical and biotechnological products are solids when sufficiently pure. As a matter of fact, careful crystallisation processes may lead to the formation of pure crystals, even in the presence of one or more additional crystallisable solutes. Crystallisation rates, however, should be carefully controlled to avoid inclusions. In some cases, contaminants may adsorb at crystal surfaces without being included at significant levels in the crystal lattice. These contaminants influence the overall purity (in the ppm range) as well as the crystal habit (shape), the crystal growth and nucleation processes. This last phenomenon has been demonstrated by various authors [23, 24]. Fractionating crystallisation techniques at low crystal growth rates, which employ reflux streams of purified product may yield extremely high product purities. An example is the so-called Thijsse wash column for melt crystallisation.

The control over supersaturation is one of the essential aspects of crystallisation. Because of the limited thermal stability of many biopharmaceutical products, evaporation of the solvent is often a less desired method since the heat transfer to the system is associated with temperature gradients. Therefore, alternative methods to remove the solvent have been proposed. One of these techniques is osmotic dewatering in which solvent removal is a pressure driven transport of solvent through solvent-selective membranes. The membrane part of the process is analogous to ultrafiltration for macromolecules or to reverse osmosis for small solutes.

Another option is extractive crystallisation. Here, the tendency of particular aqueous-solvent mixtures such as water-propanol, water-amines, water-micelles, water-polar polymers to split into two liquid phases upon small variations in temperature is used to dehydrate solutions of crystallisable solutes. At low temperatures, these systems form homogeneous mixtures, whereas at high temperatures, a solvent rich phase is created. The aqueous solute becomes concentrated in a smaller volume and consequently crystallises, whereas the pure solvent is recycled. Also, alternative schemes may be used depending on the exact phase behaviour of the component. For instance, a solute such as amino acids and peptides may crystallise from an aqueous solution upon introducing a fully miscible component, such as in water-ethanol mixtures. In a second stage, after the separation of the crystals, the conditions may be altered to induce an $L-L$ phase split that allows easy recovery of the auxiliary component. Maurer and co-workers [25] described the use of high pressure CO_2 in water-alkanol systems. At low pressures, hardly any CO_2 dissolves in the aqueous-organic mixture, but at high pressures a biphasic system is created of an apolar CO_2-alkanol-rich phase and a water-rich phase. Relatively polar solutes such as amino acids will dissolve well in the aqueous phase at high pressure and will crystallise upon releasing the pressure.

2.2.5
Membrane-Based Separations

Membranes can be characterized and classified on the basis of the applicable driving forces across membrane as well as on the phases at either side of the membrane (gas-gas, gas-liquid, liquid-1/liquid-2). Such a classification, describing most current commercial categories of membrane separations, is given by Wesselingh and Krishna [26]. Most conventional applications relevant for the biotechnology and fine-chemical industry deal with a liquid feed phase. The relatively low volatility of biomolecules in most cases often just introduces a liquid permeate flow as well. Most of these technologies – ultrafiltration, microfiltration, reversed osmosis and electrodialysis – are reasonably well described and analysed in most separation texts. In most cases, the differences in size and charge between the components in the mixture are relatively large.

A relatively novel field is nanofiltration. Nanofiltration for the separation of mixtures of structurally similar components of low and medium molecular weights is currently one of the areas in which breakthroughs in molecular selectivity would have the most impact.

When the selectivity of a single membrane in a single-stage process configuration is insufficient, multistage fractionating systems may offer a challenging technological solution. Recent successes in membrane-based fractionation technology were described by Keurentjes and Voermans [17] and Overdevest et al. [18]. They developed a multistage, counter-current fractionating system for fatty acids, and a similar four-section system using supported liquid membranes for the complete separation of enantiomers from racemic mixtures. Because transport rates of large molecules through membranes are very low, these systems do not seem particularly useful for fractionation of mixtures of large molecules such as proteins or polysaccharides.

2.2.6
Extraction

Extraction is often used in the fine-chemicals and biotechnology industry. Extraction technology has a number of distinct advantages (selectivity, capacity, robustness and good scalability), but an even longer list of disadvantages: expensive solvent recovery, many practical problems such as emulsification and the mutual miscibility of solvent and water, solvent aging by oxidation and other chemical reactions, environmental and safety aspects because of toxicity, explosivity and flammability.

Extraction is used at a large scale in carboxylic acid processes. The world market for carboxylic acids is still growing particularly for application in renewable plastics. Therefore, the need for cleaner processes that consume less auxiliary materials and produce less waste salt (gypsum) becomes stronger. Until now, alternative processes based on selective extraction of the acid from the fermentation broth by using in situ extraction, (supported liquid) membranes or electrodialysis, have not led to feasible large-scale alternatives. Various interesting approaches using pressurized carbon dioxide are used to acidify an aqueous carboxylate solution [27]. The advantage of using carbon dioxide as the acidifying agent is that it can easily be recovered by reducing pressure. A claimed advantage is the formation of (bi)carbonates that may be recycled in the process, in a dissolved or solid form. This can lead to a fully integrated process (Fig. 8) with respect to recycling auxiliary chemicals, of course at the expense of an increased energy consumption.

Aqueous two-phase technology based on polymer-polymer or polymer-salt systems may be a possible alternative to organic solvent extraction. It has the advantage that proteins and other biological macromolecules can be extracted in these systems, without loss of biological activity. Losses during polymer recycling have remained a critical bottleneck. Several alternative technologies have been developed to recycle salts efficiently, such as an extractive crystallisation procedure developed by Greve and Kula [28]. Thus far, only very few industrial applications have been demonstrated in the open literature [29].

A few interesting approaches to cope with the problem of recycling the auxiliary components in a more efficient manner have now also been proposed. These are based on relatively simple chemicals such as non-ionic surfactants [17,30] or water-soluble ethylene-propylene oxide copolymers [31]. These systems require only small amounts (several wt%) of these chemicals to produce ATPS with mi-

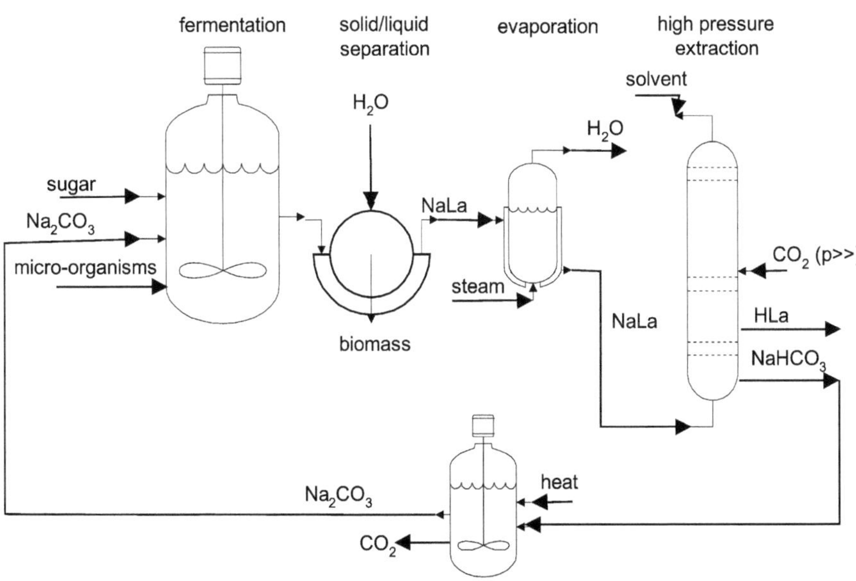

Fig. 8. Conceptual flow diagram for an integrated lactic acid production process

celle-rich and micelle-poor phases. Varying temperature leads to homogeneous solutions or to precipitates that allow relatively easy handling and recovery. They have been applied with success at the laboratory scale for the recovery of recombinant proteins [30, 31] and smaller solutes [17].

An alternative concept uses volatile compounds [32] involving combinations of NH_3/CO_2. These components form aqueous salt solutions with several ionic species such as carbamate and bicarbonate at concentrations up to 45 wt%. These solutes form aqueous two-phase systems ATPS with the usual water-soluble polymers such as poly(ethylene glycol). Because of the high ammonia content, applications are limited to pH 9–10.

2.2.7
Separation Technology for Near-Identical Particle Mixtures

The rapid developments in molecular biology have boosted expression levels in fermentations beyond the solubility of the product. This leads to solid bioproducts. The formation of solid bioproducts (crystals, precipitates) occurs either intracellularly, which requires cell disruption, or extracellularly. These particles are in the range of 1–100 μm. In other cases, parts of cells may be the desired products (membrane-bound proteins, receptors, complexed DNA). These particles are typically one or two orders of magnitude smaller (10–100 nm) and need to be recovered from streams that contain particles in the same size (and density) range. Neither conventional filtration nor centrifugation techniques are particularly suited for recovery in this size range. Also, compact biocatalytic processes may involve the conversion of suspended substrates into suspended products using immobilised biocatalysts or whole cells. These crystallisation-reaction systems are described in more detail in a later section. Some of these systems are effectively four-phase systems (S_1, S_2, S_3, L). The residence time of each solid phase must be different from the others and must be rather well controlled for proper operation (complete conversion and product separation).

Therefore, the technological challenge is large, particularly in the case of particle mixtures with near similar physical properties such as size and size distribution, density and morphology. Then, differences in surface chemistry can be exploited to separate the particles, for instance via flotation [33], $L-L$ interfacial partitioning [34–36], foam and gas aphrons (stabilised micro-bubbles) fractionation, and electrophoretic and electrostatic techniques. This whole field, despite its maturity in other industries such as metallurgy and solid waste fractionation, is totally underdeveloped for fine-chemical and biotechnological production methods.

Of particular interest is the aforementioned interfacial partitioning technology. It has been demonstrated that small particles in a mixture partition differently to the interface of a suitable liquid-liquid system, such that (1) a particle-stabilised interfacial layer develops, and (2) that particles with a 'high-affinity' displace those with lower affinities [34, 35]. This opens possibilities for the development of a particle fractionation technology to produce essentially pure particles from a mixture, as is shown schematically in Fig. 9.

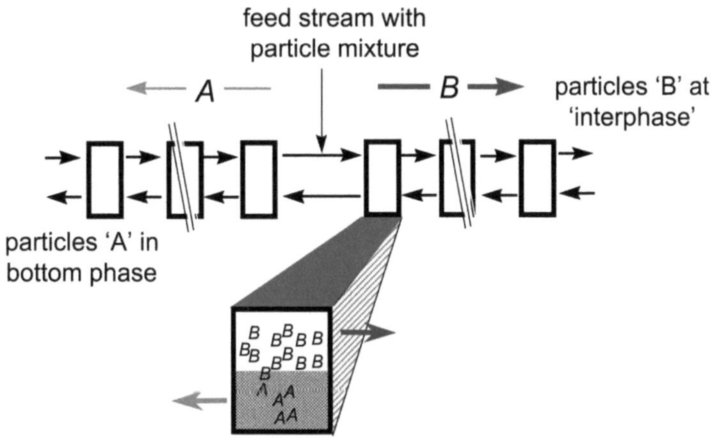

Fig. 9. Simplified schematic diagram of a particle fractionator

2.2.8
Exploiting Self-Aggregation

Many biomolecules spontaneously aggregate into micelles, gels, lamella, flocs and many other colloidal structures. The recovery may "simplify" into a simple physical separation such as decanting. Self-aggregation sometimes requires the addition of auxiliary agents such as flocculating, gelling and complexing agents. Often, complexation phenomena and, more generally, molecular recognition play an important role. This phenomenon is observed and industrially applied, for instance in producing the aspartame precursor from an L-aspartate derivative and DL-PheOMe. In this case, the remaining (undesired) D-enantiomer of phenylalanine methyl ester complexes preferentially with the wanted dipeptide product, leading to selective precipitation [37] of the complex. Larger biomolecules usually show an even richer phase behaviour that is neither well characterised nor exploited on a rational basis. A recent overview is given by Prybycien [38].

Also, auxiliary compounds can demonstrate very interesting self-aggregative behaviour, which allows controlled interaction with the desired products. We have mentioned already the example of aqueous two-phase systems on the basis of aqueous polymer-polymer, polymer-salt and surfactant-based micellar systems. Exiting developments are achieved with block copolymers composed of two alkyl chains connected by a hydrophilic polymer. Modification of the chain lengths of the blocks allows variation in the lower critical solution temperature (LCST – onset to phase separation) from 273 K to 333 K. Typically less then 5 wt% of polymer is required to construct these systems.

The partitioning of solutes in these systems is analysed in general terms by Johansson et al. [39]. It was shown that partitioning behaviour, although resulting from complex interactions, could often be correlated with fairly simple models [40]. When the problem of efficient surfactant or polymer recycling is solved, these systems may offer excellent and environmentally benign alternatives to the conventional organic solvent extraction. Additional advantages are the non-volatility, inflammability as well as the chemical and biological inertness of the polymers. These systems also have a hardly exploited potential for related techniques such as extractive crystallisation, gradient elution in liquid-liquid chromatography [41] as well as "intelligent" chromatographic resins.

2.3
Multifunctional Bioreactors

Bioconversions at an industrial scale, although highly selective, are seldom complete with respect to all substrates in a single step or pass and often require recycling of the unconverted substrates. Also, while the product in the bioreactor is just waiting for full substrate conversion, it may degrade. These two reasons are the main motives to integrate biotransformation and separation technology. The field of multifunctional bioreactors was mostly of academic interest in the past 30 years, but now seems to attract industrial interest due to its potential to enhance the performance of biocatalytic processes. This field in fact comprises three related areas: (1) integrated enzymatic reactor-separators, (2) in situ product recovery in fermentation and (3) reactive (bio-)separations. With respect to fine-chemicals production, we discuss integrated enzymatic reactor-separators only.

2.3.1
Enzymatic Bioreactor-Separators

Industrial enzymes are usually hydrolases that catalyse hydrolysis or synthesis reactions in aqueous environments. For thermodynamically controlled hydrolysis reactions, the equilibria can – in principle – be shifted completely to the product-side by dilution (increasing entropy of product formation). Thermodynamically controlled synthesis reactions using the reverse action of hydrolases can be enhanced by using excess of the cheaper reactants. This does not lead to compact processes, and affects their economic feasibility in a negative manner. Therefore, possibilities to selectively remove reaction products from each other or from the reactors during the reaction are very attractive.

The so-called crystallisation reactors are successful in the sense of being implemented at an industrial scale. Other integrated enzymatic reactor concepts that rely on the complete and selective separation of one compound from the reactor have been less successful so far. The obvious reason is that reactants and products are essentially very similar. Only when for example specific effects such as a pH-dependent charge can be exploited, one may find "simple" one-pot concepts that work. Clearly, this situation is similar to what has been observed in the previous sections on (non-reactive) separation technology, namely that more complex fractionating concepts may work, even for relatively close distribution coefficients. We will demonstrate this fractionation reactor concept for a hydrolysis using the general model as is shown in the Appendix. The reaction process as well as partitioning over the two phases is assumed to be at equilibrium in this simplified approach.

2.3.1.1
Hydrolysis Reaction

We use the hydrolysis of A into P and Q as an illustration. Examples are the hydrolysis of benzylpenicillin (pen G) or the enantioselective hydrolysis of L-acetyl amino acids in a DL-mixture, which yields an enantiomerically pure L amino acid as well as the unhydrolysed D-acetyl amino acid. In concentrated solutions these hydrolysis reactions are incomplete due to the reaction equilibrium. It is evident that for an accurate analysis of weak electrolyte systems, the association-dissociation reactions and the related phase behaviour of the reacting species must be accounted for precisely in the model [42, 43]. We have simplified this example to neutral species A, P and Q. The distribution coefficients are $K_Q=0.5$ and $K_P=K_A=2$. The equilibrium constant for the reaction $K_r=x_p x_Q/x_A=0.01$, where x is a measure for concentration (mass or mole fractions) compatible with the partition coefficients. The mole fraction of A in the feed (z_A) was 0.1, which corresponds to a very high aqueous feed concentration of approximately 5 M. We have simulated the hydrolysis conversion in the fractionating reactor with 50–100 equilibrium stages. A further increase in the number of stages did not improve the conversion or selectivity to a significant extent. Depending on the initial estimate, the calculation requires typically less than five iterations.

A typical concentration profile for a 50-stage fractionating reactor, with a feed at stage $N_F=35$ is given in Fig. 10. V runs from top (stage 1) to the bottom (stage 50) and L in the opposite direction. The feed was concentrated $z_A=0.1$, and the flow rate ratios $m_1=1.667$ and $m_2=1.833$. The conversion under these conditions was 90.3% (in batch 10.5%), with purities for Q of 84.8%, and for P of 94.83%. Further diluting the feed stream increases the conversion further. Because the partition coefficients of A and P are equal, their separation factors are also equal and they move in the same direction (towards the bottom section of the reactor with V). This leads to the parallel concentrations profiles of A and P in Fig. 10.

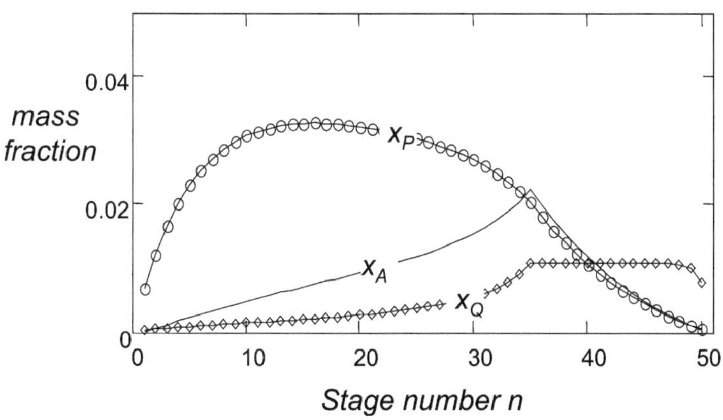

Fig. 10. Calculated composition profiles in a fractionating reactor. The conditions are shown in the text

Varying the flow rate ratios in a systematic manner gives an insight into optimal conditions. The results are summarized in Fig. 11 (conversion) and in Fig. 12a, b (purities of P and Q). The conversion increases close to the diagonal. This is partially a dilution effect: $m_2 - m_1 = F/V$ decreases to small numbers while approaching the diagonal. Figure 12a and b show the variation of purity of the products Q (12a) and P (12b) respectively, while varying (m_1, m_2) approximately parallel to the diagonal. The closer the flow rate ratios are to the distribution coefficient of a species, the better the criterion for the 'other' species is satisfied. This results in an increased removal of the 'other' species and thus a higher purity.

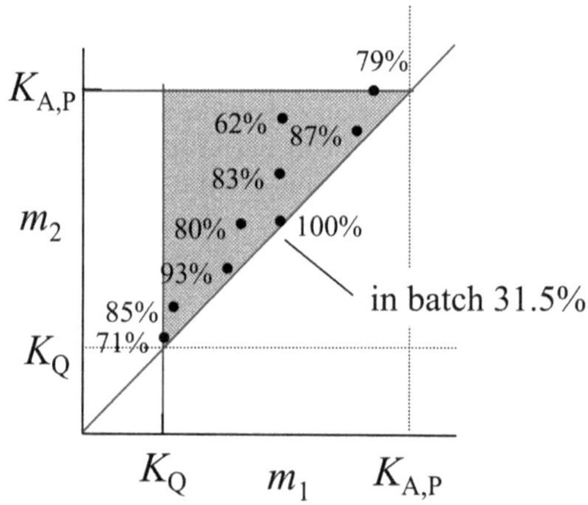

Fig. 11. Conversion (degree of hydrolysis) in a fractionating reactor. *Dots* represent (m_1, m_2), the corresponding *number* is the calculated conversion for data in the text. The batch conversion, corresponding to the 100% conversion point, is limited to 31.5%

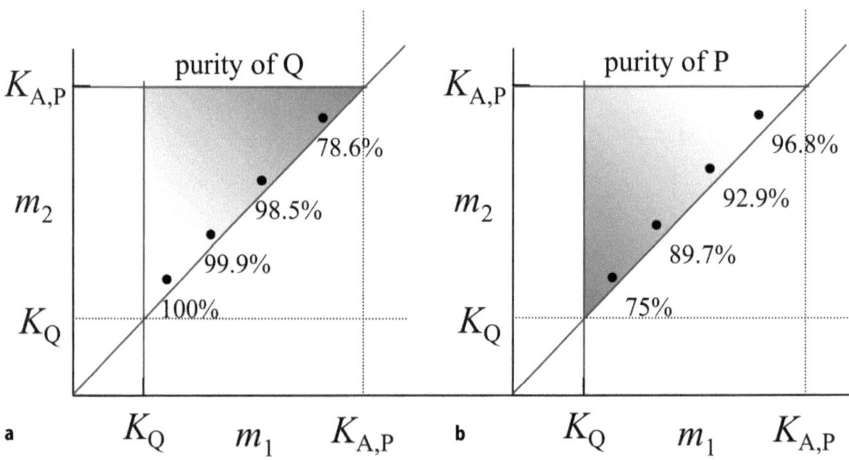

Fig. 12 a, b. Purities of **a** product Q (*left panel*) and **b** product P (*right panel*) for varying (m_1, m_2)

This example is a worse case analysis. Systems with (1) a more dilute feed, (2) with $K_Q < K_A < K_P$ and (3) in which the partition coefficients are more different, can lead to complete conversion in a few stages. This is investigated in great detail by Den Hollander et al. [14–16]. Also, manipulating the local partition coefficients in different stages by varying pH, salt concentrations or solvent composition, offers a large potential for further optimisation. A last area that is practically unexplored is to use internal recycle streams (refluxes), which can lead to accumulation of specific products in specific sections.

2.3.1.2
Fractionating Synthesis Reactor

At this moment, fractionating reactors are mostly studied and applied outside the fine-chemical field. Examples are the large-scale production of the fuel ethers MTBE and TAME via reactive distillation. Also, biocatalytic studies have been performed. Malcata and co-workers investigated the integration of ester formation by lipases and distillative separation of the final products ester and water [44]. A number of synthesis reactions have been studied such as the esterification of ethanol and acetic acid to form ethyl acetate and water [45] in an SMB reactor with chemocatalysts (acidic ion exchange resins). Another, fairly similar application was presented by Kawase et al. [46] to manufacture an ester from 2-phenylethanol. Mensah and Carta [47] used a chromatography column with lipases immobilised on resin to produce esters as well.

2.4
Rational Design of Integrated Processes

It is evident that many alternative process concepts, differing widely in process conditions and feed stocks, can lead to the desired product. A quantitative comparison of these alternatives is required, which asks in its turn for quantification of molecular properties and operating conditions. Thus, selection and rational design greatly benefit from the availability of reliable thermodynamic data as well as predictive models.

2.4.1
Thermodynamic Models

Intuitive qualitative concepts based on substantial empirical experience such as "hydrophobicity", are being used to quantify and rank molecular properties and operating conditions of processes. Identifying the underlying, general and quantitative relations to thermodynamic properties can assist in translating this valuable knowledge into quantitative tools, such as computerized models. Gude et al. [48] as well as Van der Wielen and Rudolph [40], developed a general methodology to correlate limiting thermodynamic properties which are of use in a wide variety of existing separation processes, including crystallization, aqueous-organic and ATPS-extraction, ion exchange, sorption and membrane processes. It was shown that the parameters in this general methodology can be obtained from a limited number of experiments, translated across the boundaries of different separation techniques and be predicted from data that commonly characterize the final products. It was demonstrated that this approach helps in developing quantitative insight into complex heterogeneous systems such as CO_2-aided extraction with organic solvents. This work focused on small biomolecules that could carry several charges and be overall neutral (zwitterionic species) but also might have a net charge (ions). Typical classes of molecules are amino acids, various β-lactam antibiotics and small peptides. Such a general thermodynamic framework also allows in principle the extension to other classes of biomolecules, as was demonstrated by Johansson et al. [39] with a Flory-Huggins based model.

2.4.2 *High-Speed Experimentation*

Collecting reliable thermodynamic data has always been a tedious and laborious activity. This situation is anticipated to change soon. The development of miniaturised, array-based high-speed screening techniques in combination with combinatorial (bio-)chemistry has already yielded excellent result in the development of affinity ligands for chromatographic resins. For instance, libraries of monoclonal antibodies, phages and dyes have become available commercially and are extensively used in the development of specific costumer tailored resins. It is now a matter of time (and money) to generate and exploit similar libraries for screening other auxiliary compounds as well as to characterise the thermodynamic properties of large groups of bioproducts for instance while simultaneously screening for a particular drug or active ingredient. Although the high-speed experimentation (HSE) methodology seems potentially able to reduce the experimental costs greatly, reliable model-based predictions can prove alternative and complimentary pathways and assist in obtaining rapid insight into feasible mechanisms.

These methods may also be used to – in silico – generate new, optimised molecular structures that are as yet difficult or even impossible to synthesize. In this manner, molecular computations may generate a driving force for novel chemistry. Of particular interest in this respect are molecular dynamics simulations that enable a quantitative description of transport rates of solutes in porous structures. There is only qualitative insight into mechanisms that may be exploited to separate compounds via the topology of the internal structure and chemical compositions of separations media such as adsorbents and membranes. We anticipate that both experimental and computational approaches when integrated will lead to high-speed process development.

2.4.3
Tools for Analysis and Design of Complete Processes

2.4.3.1
Starting Points for Process Design

In (fine) chemical processes, matter and energy streams are converted into valuable, sometimes structured products. New processes for new products in this sector are often based on chemists' insight and developed along 'chemical methods', often at a laboratory bench. The use of biotechnology methods changed and extended the possible set of chemical tools and methods, removed old constraints and added some new ones. But only minor progress was made in the development of rational and generic methodologies for the conceptual design of fine chemicals processes in "green field" or "grass root" situations[2]. Most industrial fine-chemical processes are still in essence geometrically scaled-up versions of the laboratory bench systems. Process flows are also essentially linearly scaled-up and no positive scale effects seem to have been obtained.

In general, process design comprises a sequence of development steps: defining the 'process', generating process alternatives, and evaluating and optimising them for particular situations. In the first stage of process design, the 'process' must be defined in terms of specifications for the product (composition, structure and function) and other chemical components, in terms of plant site, market, and in terms of environmental, legal and safety constraints.

An obvious second point in process design is the economic potential. This is the price difference between final products and raw materials or intermediates, at stoichiometric or realistic yield conditions. Positive values indicate a potentially interesting candidate feedstock. Since prices are not absolute measures and fluctuate in time, a scenario analysis should be included as well. Alternatively, the economic potential may indicate which minimal overall yield should be obtained to achieve certain margin targets, and which challenges technology development has to meet.

2.4.3.2
Feasibility of Process Alternatives

In this stage, technological tools become more important. The 'soft' information or knowledge available for the crucial first steps in process design deals with known (and to-be-discovered) chemical and physical phenomena of the components involved such as phase changes, reactions and transport phenomena. The corresponding 'hard' quantitative information required to estimate the theoretical feasibility of particular process steps is represented by the thermodynamic properties of these components. For a particular system, this allows direct calculation of the absolute criterion of feasibility: second law of thermodynamics ($\Delta G = 0$).

---

[2] In which case, no base case process or other 'prior art' exists.

We can see, however, three main problems in this stage: (1) the availability of reliable, quantitative thermodynamic data, (2) the practical feasibility of particular process steps and (3) the step-up from the feasibility of individual phenomena to that of integrated systems.

The first aspect, related to measuring and predicting thermodynamic data for fine-chemicals processes, is gradually attracting more attention. It is evident that high-speed experimentation (HSE) methods based on micro-arraying and other miniaturisation techniques can dramatically increase the throughput and volume of experimental work. This development may lead to important leaps in filling-in databases. It will also accelerate the generation and testing of improved mathematical models for the prediction of these properties. Reliable predictive models can reduce the necessity for experimental tasks (and time) significantly [40,49].

Solving the second aspect, practical feasibility, is more troublesome. Practical feasibility relates to experience and insight obtained in existing plants or earlier process development projects, under similar specifications and constraints. This experience is usually within human beings, and often in an implicit form. It is therefore difficult to extract and reshape into a set of qualitative or quantitative rules.

Problems with the practical feasibility of alternatives for existing processes can often be attributed to undetected deviations from earlier implementations. Several approaches are known that may assist in using existing experience in the design of a new process. As an example, Asenjo and co-workers [50] proposed a diagnostic expert system based on a commercial expert shell and an experimental database of the selected properties of main contaminants in microbial production processes. Rules were developed on the basis of the experience of many industrial process designers, basically summarising the state-of-the-art at the time of the questionnaire. Using relevant databases and cost functions, the computer model seems capable of generating realistic alternative process sequences of unit operations for biopharmaceutical production.

However, problems with the practical feasibility of "green field" processes for novel products, can also relate to the poorly understood behaviour of components, to overlooked details in equipment design, to interfacing problems of unit operations, and to insufficient insight into the systems behaviour as a whole. This class of problem is particularly difficult to predict, or even detect at a laboratory or pilot scale level. To some extent, rigorous modelling methods can be used for scale-up problems (CFD for flow problems, finite element methods for mechanical problems) or to analyse system behaviour (reactor or separator equipment model, flow sheeting software). Unfortunately, the underlying (thermodynamic) models for components are still insufficiently accurate, and the composition of process flows in realistic fine-chemicals processes is subject to variation. Predicted results are often insufficiently reliable.

The third and last class of problem for process development originates from the fact that many isolated physicochemical phenomena do not occur spontaneously. Very similar to processes in living systems, energetically unfavourable phenomena can be driven towards completion by (thermodynamically) coupling them to other phenomena. Although this concept of linear energy converters is fairly common at the microscopic level in the metabolism and transport in living cells, it is not general at the macroscopic level of fine-chemicals plants. Examples are integrated reactor-separator systems (sorptive, extractive and membrane reactors), such as described elsewhere in this work. It remains, however, challenging to find a working set of complementary processes.

Again, calculating the ΔG of the whole system seems to be a good qualitative measure for theoretical feasibility. It should be remembered, that coupling phenomena leads in general to an increased inflexibility and sometimes also to highly unexpected non-linear systems' behaviour (impossible to start up/close down, multiple or cyclic steady states, run-aways).

2.4.3.3
Process Efficiency

The above technological tools to aid process design indicate feasibility only (that is "can it be done?"). They do not compare process alternatives by a generic measure for efficiency (that is "how well can it be done?"). Again, a thermodynamic starting point can be taken to obtain a quantitative measure of process efficiency.

The second law of thermodynamics dictates that all real processes inevitably lead to entropy production or, formulated differently, to a lower energetic quality of the product flows compared to the input flows [51]. Let us analyse Escher's "Waterval" (1961) in which a perpetual flow of water drives a hidden black-box process. When the absurd part of the process is removed, the common schematics of a real process are obtained, as shown in Fig. 13. The water flow represents the work (in a thermodynamic sense), necessary to perform this specific process.

The minimum reversible work requirement of a separation process is solely given by the composition and conditions (T, P) of the feed and product streams [52]. It can be calculated from the difference in Gibbs energy of product and feed

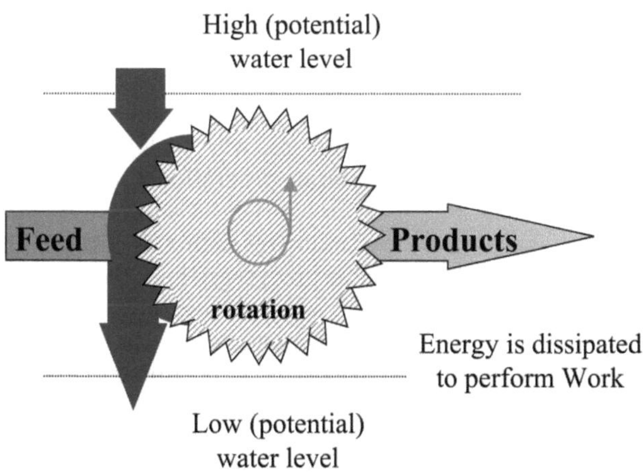

Fig. 13. Schematics of Escher's "Waterval" (1961), representing a real process

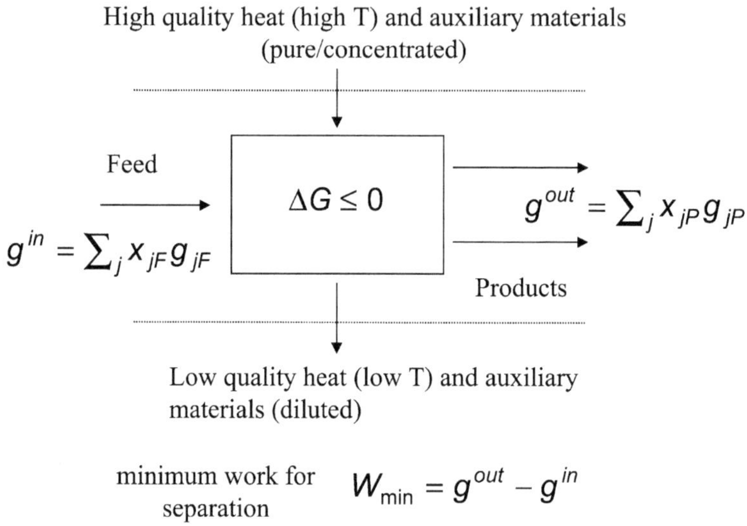

Fig. 14. Processes as open systems, driven by the input of heat, mechanical work and auxiliary materials [52]

flows. This is shown in Fig. 14 for an ideal binary separation. The work can be performed on the system in terms of mechanical work, heat energy or material flows.

A more detailed analysis of various bioprocesses indicates that consumption of auxiliary materials is a main contribution to the work input of bioseparation processes. This can be expressed in terms of Sheldon's *EQ*-factor [53] as well. The *EQ*-factor is the product of the environmental coefficient (kg of waste per kg of product) and a weighing factor *Q*, which indicates the quality of the waste; this ranks waste from harmless (low *Q*) to highly toxic (high *Q*).

In real (fine) chemical processes, concentrated materials are mixed at great exergy loss in huge quantities of water and other solvents. The problems created here have to be solved in the downstream processing. The recovery and purification of the desired product demands a further work input in the sense of 'mixing' the feed with (pure) solvents (precipitation and extraction), salts (ion exchange), heat (evaporation and solvent recovery), electrical power (electrodialysis), pressure (filtration and membrane separations) or just extra water (gel filtration).

Thus far, we have discussed the minimum, reversible work requirement (which is only valid for infinitely slow, reversible processes). Real processes, however, are operated at a finite rate and under irreversible conditions. This leads to additional friction (leading to energy dissipation), which has to be balanced by extra work input. For instance, we state that useful work is proportional to the flux *N* (or rate) of a species through the process, and will be approximately proportional to its driving force. The driving force is given in Fig. 15 as a chemical potential gradient.

Lost work, however, is given by the product of flux *N* and driving force, and is therefore proportional to the driving force squared. At low driving force, only

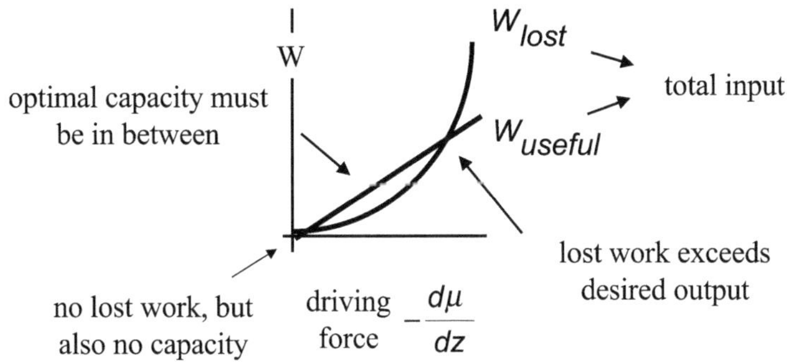

Fig. 15. Useful and real work requirements as a function of the driving force of the process (here: chemical potential gradient)

small amounts of work are lost, but the capacity of the process also is low, which is undesired. At high driving forces, however, lost work (proportional to driving force squared) may well exceed useful work. Operation at intermediate driving force appears attractive to optimise the ratio of useful work and lost work. This is also demonstrated in Fig. 15.

A more generic approach to quality analysis of integrated processes quantifies the energetic quality of a process stream in terms of exergy [54]. Exergy is the (remaining) Gibbs free energy which *can* still be extracted from the system. Probably the most beautiful feature of exergy is the unified description of the quality loss of these streams in terms of kJ mol^{-1}. This provides a unified basis for comparison of fairly different process set-ups. This is not possible with other indicators for process quality such as heat consumption or Sheldon's *EQ*-factor [53].

3
Case Study: Semi-Synthetic Antibiotics (SSAs)

The industrial manufacture of semi-synthetic penicillins and cephalosporins is an outstanding example of the integration of chemistry and biocatalysis. The impact of biocatalysis shortens the synthesis for Cefalexin from ten to six steps is a successful example (Fig. 16) [55, 56].

In the crucial final step in the Cefalexin synthesis, the cephalosporin nucleus 7-ADCA is coupled with phenylglycine amide or ester. This is one of the first industrial examples of a synthesis reaction performed by enzymes. Until then, enzymes were mainly employed for hydrolysis; the deacylation of penicillin G to give 6-APA (not shown), that of cephalosporin G to give 7-ADCA (Fig. 16) as well as the kinetic resolution of the DL-phenylglycine derivatives (Fig. 16) are examples. Also, similar processes were developed for other semi-synthetic antibiotics derived from phenylglycine and 4-hydroxyphenylglycine (Fig. 17).

From an environmental point of view, these processes are very beneficial because of the elimination of halogenated solvents and several reagents and the re-

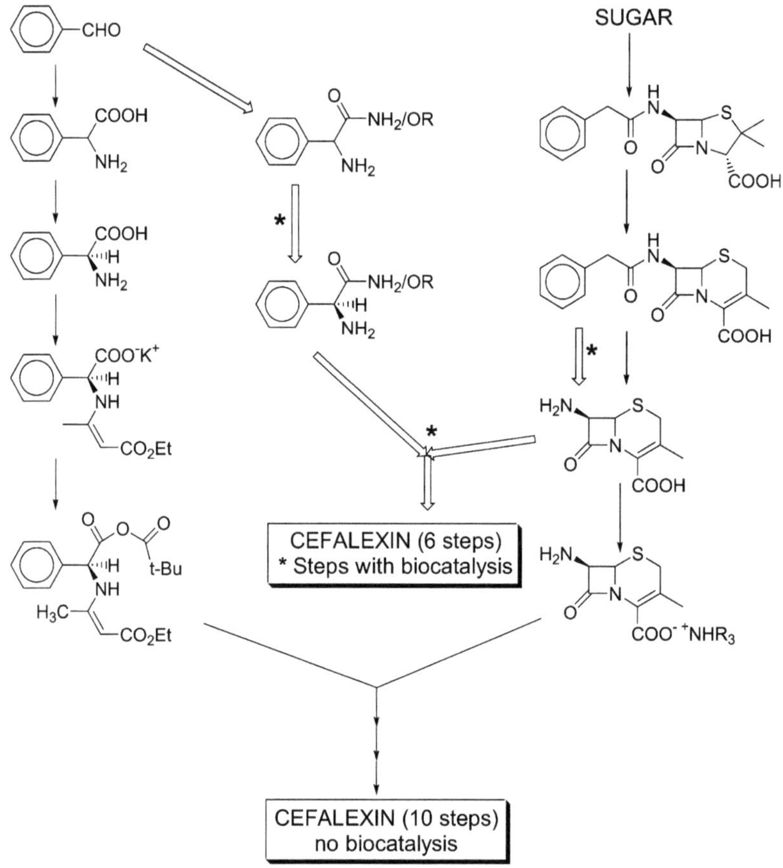

Fig. 16. Traditional (*single arrow*) and modern (*double arrow*) biocatalytic process for Cefalexin.
* Indicate biocatalytic steps

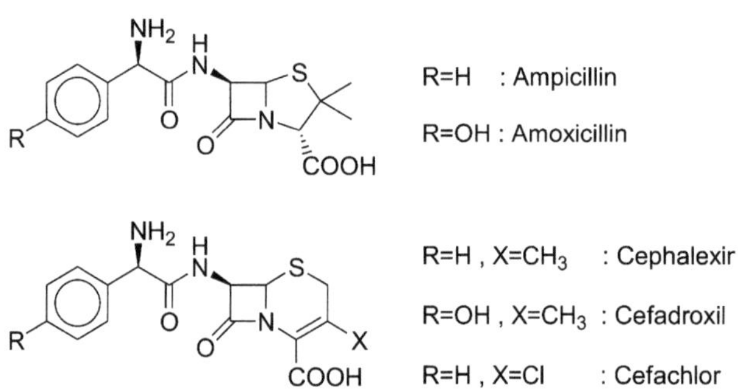

Fig. 17. Penicillins and cephalosporins for which enzymatic coupling processes have been developed

duction of waste streams of inorganic salts. Expressed as kg of waste per kg of product a reduction of 30/1 to 5/1 has been achieved [2].

A common bottleneck in these processes remains the undesired enzymatic hydrolysis of the activated side-chain molecule to the usually poorly soluble amino acid [57]. In combination with unfavourable equilibrium conditions in the coupling reaction, this still leads to a tedious and costly purification technology.

3.1 Ongoing Greening

Despite many resistance problems, it is anticipated that penicillins and cephalosporins will remain prominent antibacterial drugs for another 10–20 years. Therefore, further simplifications and efficiency improvements of the manufacturing process have been investigated. A collaborative research program at several Dutch universities and DSM Life Science Products focuses at improving enzymatic processes, development of new biocatalysts, as well as further integration of chemical synthesis and biocatalysis, alternative process technologies and efficient separation technology. Several approaches and results from this program are presented below.

3.1.1 *Fermentation of 7-ADCA*

The multistep chemical conversion of penicillin G to 7-ADCA (Fig. 16) has recently been replaced by a 2-step biosynthesis (Fig. 18). This is a major step forward to shorten the industrial synthesis of cephalosporins and this has already been implemented. The 7-*N*-adipoyl-ADCA is obtained directly through fermentation with a modified *Penicilium chrysogenum* followed by a simple enzymatic removal of the amino substituent. Again, various reagents such as silylating agents, phosphor halides, pyridine and DMF, and some halogenated solvents have been replaced by biocatalysts in aqueous medium.

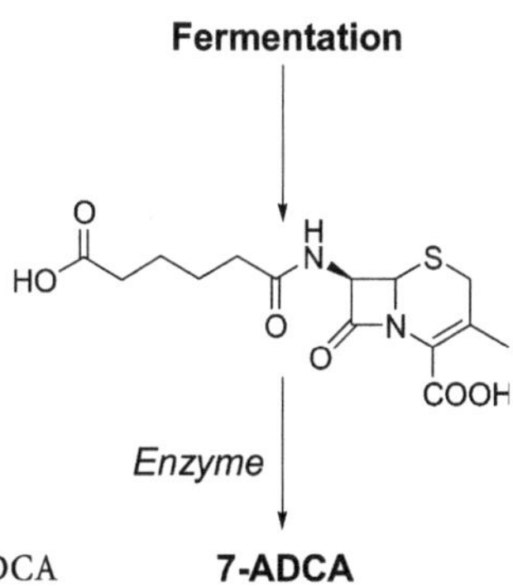

Fig. 18. Biosynthesis of 7-ADCA

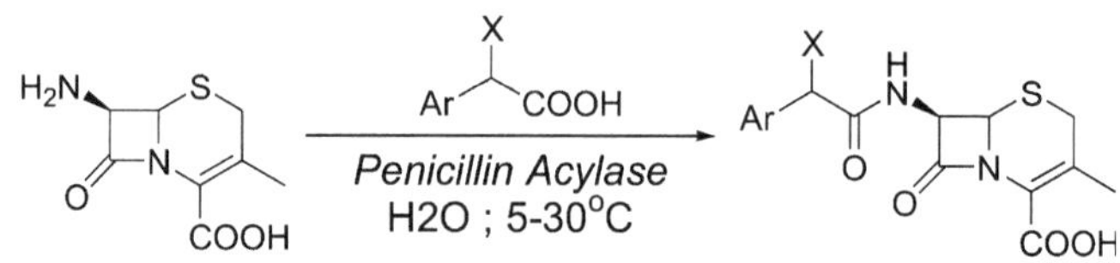

Ar	X	Coupling %	pH
Ph	H	28	6.0
3-Indole	H	4	5.0
Ph	Br	27	6.0
Ph	OH	9	5.0
4-HO-Ph	OH	6	6.0
Ph	NH_2	-	4-7
4-HO-Ph	NH_2	-	4-7

Fig. 19. Thermodynamic coupling towards β-lactam antibiotics

3.1.2
Thermodynamic Coupling

An early goal in the research program was the thermodynamically controlled direct coupling of the free side-chain amino acid with the underivatised nucleus 7-ADCA (for cephalosporins; see Fig. 19) or 6-APA (for penicillins). It has been shown [58–60] that thermodynamic coupling can be done, provided the side chain does not contain an α-amino substituent. Obviously, the zwitterionic character of α-amino acids constitutes an energy minimum, which brings them out of reach for activation towards coupling in aqueous media. Some coupling activity could be detected on replacing water by polar, hydrophilic solvents such as glycols and glymes. Conditions, however, are rather remote from industrial relevance.

Surprisingly and interestingly, several patents (French Patent 2014689, 1968; WO 91/09136, 1991) claim enzymatic coupling with a simple phenylglycine salt in water or with amino acid side chains. All of these are without proof, and are very questionable from a theoretical point of view.

3.1.3
Suspension Reactors

An effective manner to reduce reactor volume is feeding solid substrates under conditions that the products are solids as well. It has been shown that yields can at least be similar to those in conventional (dissolved product) enzymatic reactions. Suspension-to-suspension conversions are especially advantageous when hydrolytic reactions are to be reversed or suppressed. An additional advantage is that sensitive products are usually protected from degradation by occurring in the crystal form.

In these so-called suspension-to-suspension processes, simultaneous dissolution, crystallisation, and enzymatic reaction take place. In case of weak-electrolyte reactants, all sub-processes have pH effects. They influence and are influenced by the pH. Therefore, a thorough understanding of these different sub-processes is necessary for optimising most suspension-to-suspension processes. An industrially relevant and interesting example is the kinetically controlled synthesis of amoxicillin (Amox) from D-*p*-hydroxyphenylglycine methyl ester (HPGM) and 6-aminopenicillanic acid (APA). In this case, pH control can be omitted. The enzyme penicillin acylase catalyses the synthesis (reaction I), by coupling HPGM and APA. In a batch reactor, both substrates may initially be mostly undissolved, whereas most of the amoxicillin will crystallise during its production.

$$\text{I Synthesis:} \qquad \text{APA} + \text{HPGM} \rightarrow \text{Amox} + \text{MeOH}$$

The enzyme also catalyses the undesired substrate hydrolysis (of HPGM, reaction II) and product hydrolysis (of Amox, reaction III). Both side-reactions lead to hydroxyphenylglycine (HPG).

$$\text{II Substrate hydrolysis:} \quad \text{HPGM} + H_2O \rightarrow \text{HPG} + \text{MeOH}$$

$$\text{III Product Hydrolysis:} \quad \text{Amox} + H_2O \rightarrow \text{HPG} + \text{APA}$$

Integrating models for the sub-processes, can lead to a quantitative model for the complete process [61]. The model can describe the solid-to-solid reaction fairly well and can explain pH shifts during the suspension-to-suspension reaction. The model can be used to find the optimal conditions to produce Amox. For example, when the enzyme stability or activity is low in a certain pH range the model can predict whether or when the pH will be in that range and pH control is necessary. In this way no unnecessary buffers, acids or bases are used for pH control, which can simplify downstream processing. The model can also predict when to stop the reaction to achieve the highest yield of product.

3.1.4 *Product-Specific Complex Formation*

Scientists at Eli Lilly discovered the specific complexation of cephalosporins by the complexing agent β-naphthol. It has been applied to improve the yield of the enzymatic coupling to produce Cefalexin by NOVO. This process has been developed further at DSM in collaboration with the University of Nijmegen. It was shown that the β-naphthol complexation of Cefalexin brings the coupling equilibrium to near-completion. Surprisingly, β-naphthol only slightly inhibits the enzymatic coupling reaction. Also in this case, the undesired hydrolysis of the side-chain precursor remains with the implications discussed before. In addition, the efficient removal of β-naphthol is of a critical importance to meet all quality requirements of the bulk medicinal end product.

To find more environmentally compatible alternatives for β-naphthol, crystal structures of the β-naphthol complexes were determined and several other aromatics were tested in complex formation [62]. A range of complexing agents was shown to be effective complexants. Although various crystal structure types are found, the common feature is a cage formed by four cephalosporin molecules. The cage is filled with mostly two host molecules and a varying number of water molecules to reach maximum crystal lattice stability. The flexibility of the cage combined with the employment of water as cement allows for the large number of hosts that can be accommodated. The results can be used in a predictive model to develop product-specific complexation and product isolation [63].

3.1.5 *Fractionating Reactor for the Hydrolysis of Pen G*

Den Hollander et al. [14, 16] investigated the enzymatic hydrolysis of penicillin G to phenylacetic acid and 6-aminopenicillanic acid in biphasic aqueous-organic systems without pH-control. In a preliminary study, the two phases were counter-currently contacted in a discrete manner, so that equilibrium was reached in each stage. Sets of three and five shake flasks served to mimic equilibrium stages in the counter-current set-up. It was shown, that counter-current contact leads to significant improvement of the equilibrium conversion relative to the batch or co-current situation. When penicillin G was fed in an intermediate stage, either exit contained mainly one of the two products. This simplifies product recovery.

A mathematical model was used to calculate the concentrations of all components and the pH at every equilibrium stage. The pH and concentrations of the components at every equilibrium stage were predicted with reasonable accuracy. This model is based on dissociation and reaction equilibria of the compounds, stoichiometric balances and an electroneutrality equation. Precipitation of 6-aminopenicillanic acid, which was observed at a combination of low pH and high 6-APA concentration in the aqueous phase, is not taken into account in the model.

Experimental conversions in this simple system without control of pH etc. could be as high as 98%, depending on the flow rate ratios. The conversion was typically 10–30% larger in the 3-stage and over 50% larger in the (simulated) counter-current system relative to batchwise conversion. A further increase in the number of stages seems attractive, but it can be demonstrated that adding stages to systems containing over 25–50 equilibrium stages does not notably improve the conversion.

On the basis of these results, a counter-current fractionating $L-L$ reactor system with an increased number of stages is investigated by modelling. An idealized reactor of 25 stages, equipped with an axial pH-control system and operated at pH 6, would lead to the following composition profiles with near-complete conversion and purification as is shown in Fig. 20. Pen G only occurs around the feed point. PAA is transported towards the solvent outlet at stage 1, and 6-APA is transported to the solvent inlet at stage 25. A further optimisation, for instance with respect to minimal flow of solvents, feed location and effect of crystallisation, is not done here, but is the subject of future work.

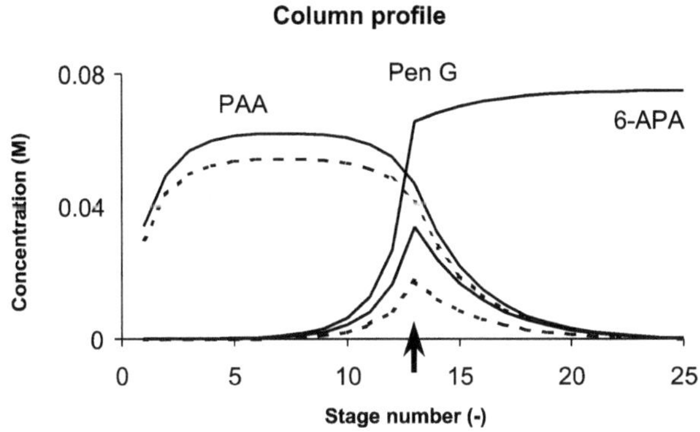

Fig. 20. Composition profiles in the aqueous (*solid curve*) and solvent phase (*dotted curve*) in a fractionating enzyme reactor for the Pen G hydrolysis

3.2
Biocatalyst Development

The fast acceptance of biocatalysis by the (fine-)chemical industry will continue to trigger a great deal of research in the areas of organic chemistry, bio-synthesis and process technology. At the same time, a lot of additional fundamental insight, that is in enzyme action, molecular biology of micro-organisms and biocatalyst formulation, is required to allow further industrial exploitation. A few challenges, both from a scientific and an applied point of view are shown below.

Even today's organic syntheses are still mainly governed by a step-by-step approach; bond cleavage and bond making are done one by one. Lack of selectivity and/or incompatible reaction conditions are the underlying causes. The high selectivity that enzymes show under comparable conditions, that is in aqueous systems, allows in principle the use of several biocatalysts in one reactor system. This could be a batch reactor, series of columns or any other system. A promis-

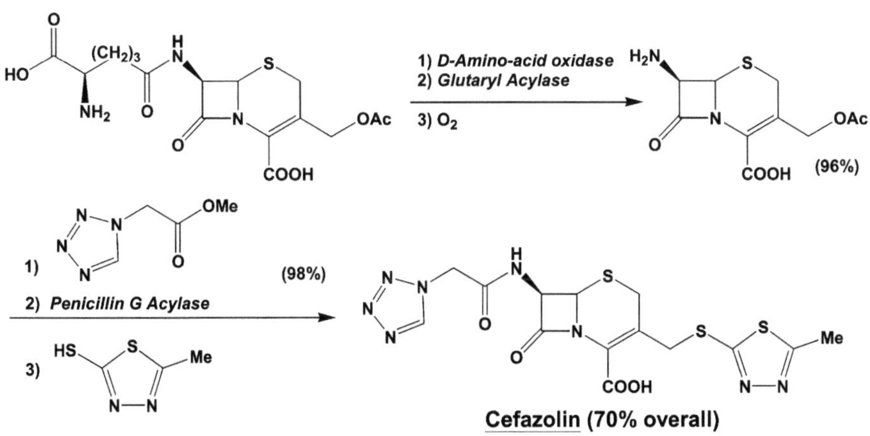

Fig. 21. Cascade catalysis in Cefazolin synthesis

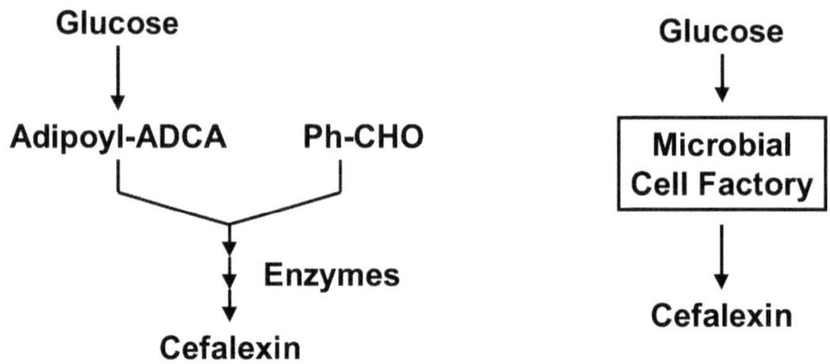

Fig. 22. Cascade catalysis and direct fermentation of Cefalexin

ing example is shown in Fig. 21 employing three enzymes and a consecutive substitution in one pot to give Cefazolin.

A challenging extension would be the introduction of those enzymes in the micro-organism employed in the fermentation of the starting material Cephalosporin C and thus allowing direct fermentation. Similar approaches can be envisaged for other penicillins and cephalosporins as is outlined in Fig. 22.

However, many problems have to be solved at the molecular biology level, before industrial application will be feasible. Transport mechanisms in micro-organisms and interaction of primary and secondary metabolism are just a few.

4 Outlook

In this work, we have – by no means completely – indicated the set of tools available to generate solutions for the development of improved and more competitive

Conversion related:

- relief of product inhibition
- circumventing the thermodynamic limit of conversion
- manupilation of metabolic control mechanisms
- improvement of selectivity towards desired product

Techno-economic:

- reduction of the number of unit operations
- reduction of process streams
- improved control through decoupling product formation and withdrawal

Fig. 23. Motives for process integration

processes for the fine-chemicals industry. The starting point for analysis of a particular process can be the set of general motives presented in Fig. 23, which outlines the most common reasons for process integration. Clearly solutions must be tuned to the requirements of particular products/processes as was shown in the case study on SSA. It is also evident that the more interesting and challenging solutions are generated by the combination of rational analysis, skilled use of theoretical and experimental tools and the open eye for creative moments.

Appendix: A Design of Non-Reactive and Reactive Fractionating Systems

This concerns essentially steps 2 and 4 of the design procedure from Table 2 and can be restricted to the basic fractionating configuration in Fig. 5. In the following, we will approximate each of the counter-current sections by a cascade of equilibrium stages, as is shown in Fig. A-1.

The feed stream is supposed to contain the same phase as L. Therefore, upon crossing the node between section I and II, the magnitude of stream L changes due to the introduction of the feed. For instance, in the fractionating extraction as shown in Fig. A-1, $V_1 = V_2$, but $L_2 = L_1 + F$. Step 2 of the design procedure requires the identification of adequate ranges of separation factors for each of the (groups of) species to be separated and for each of the sections. The (group of) substances A with the larger distribution coefficients are likely to 'move' with the stream of V, whereas the substances B with the smaller partition coefficients re-

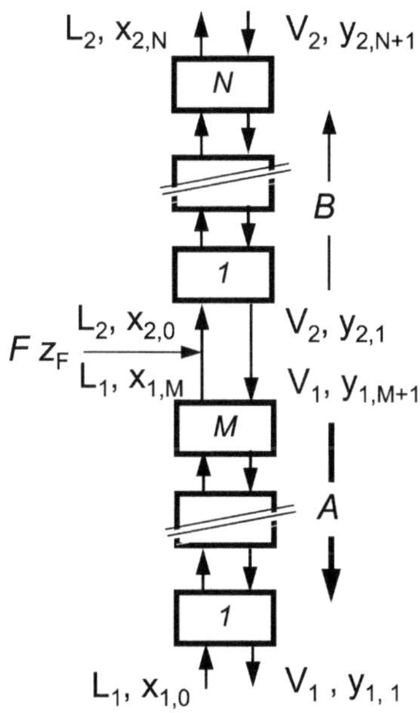

Fig. A-1. Scheme of a multistage fractionation cascade. *Arrows* determine the direction of motion of the species *A* and *B*

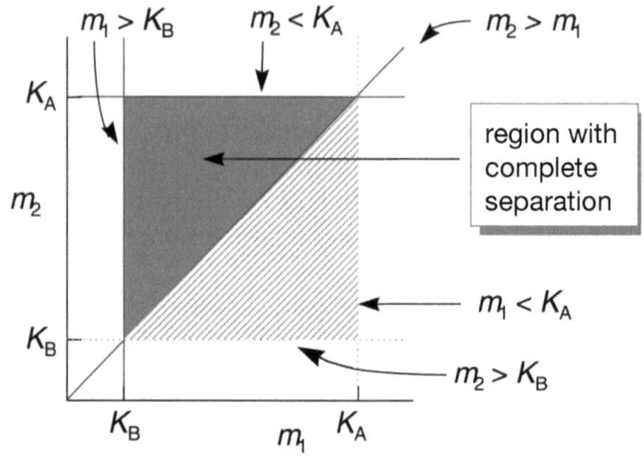

Fig. A-2. Operating conditions for complete recovery of *A* (*'heavy key'*) and B (*'light key'*) products by a two-section fractionating separation

main in the feed stream. The separation factors of A in both sections I and II should exceed unity (S^{I}_{A}, $S^{I}_{A}>1$), whereas the separation factors of B should be smaller than unity (S_{B}, $S^{I}_{B}<1$).

$$S^{I}_{A}>1,\; S^{II}_{A}>1,\; S^{II}_{B}<1,\quad \text{and}\; S^{II}_{B}<1 \qquad \text{(eq. 1)}$$

or

$$K_B<m_1<m_2<K_A \qquad \text{(eq. 2)}$$

where $m_j=L_j/V_j$. These criteria limit the operating conditions to the dark shaded area in the m_1-m_2 plane that is shown in Fig. A-2.

Furthermore, $L_2=L_1+F$. Demanding a positive feed flow ($F>0$), sets another criterion: $m_2>m_1$. The limiting condition ($m_2=m_1$) corresponds to the diagonal in the m_1-m_2 plane. The last criterion therefore limits relevant operating conditions to the shaded, triangular upper-diagonal area.

Any point in the triangle can result in complete separation of a mixture of components A and B, as well as in complete recovery of each individual component (A in the *V*-stream and B and the *L*-stream), provided that the sections contain sufficient numbers of the equilibrium stages. When the distribution coefficients of A and B are very similar, and a low value of the stream of auxiliary material is aimed at, substantial numbers of equilibrium stages are necessary. This is, for instance, the case for the chiral separation of racemic mixtures into pure enantiomers using counter-current chromatography. This diagram was first constructed by Morbidelli and co-workers [64], and can be generalized for any type of counter-current separation system [13].

The optimal point for efficient usage of auxiliary material in the *V*-stream is represented by the upper-left corner of the triangle: $m_2-m_1=F/V$ is largest at this point. Robust operation is effected by allowing a larger consumption of *V*, in accordance with expected fluctuations in process operation (for instance by operating at a 10–30% higher consumption).

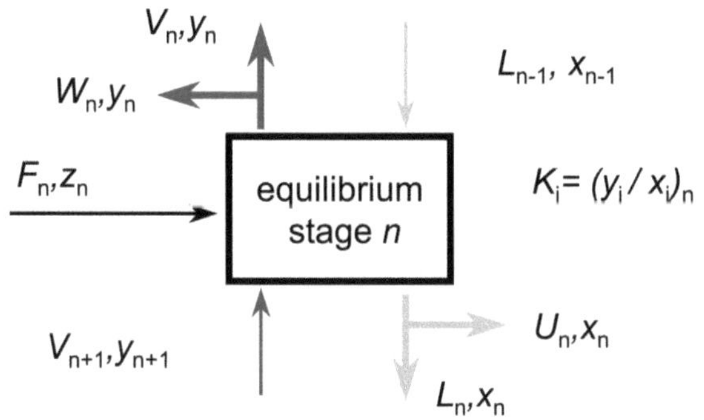

Fig. A-3. Schematic representation of a non-reactive equilibrium stage with a feed (*F*) and side withdrawal streams (*U*, *W*)

Non-Reactive Fractionators

A general procedure to describe fractionating contactors is by assuming a cascade of interconnected stages, numbered from 1 (top) to *N* (bottom). A typical stage is shown schematically in Fig. A-3. The stages can – in principle – have a feed stream *F* and withdrawal streams (*U*, *W*). A stage may be an actual tray (distillation, extraction) or be a theoretical tray, representing a certain length of bed. A feed stream at one stage may be an external feed, but may also include internal streams, withdrawn from other stages. This allows recycle flows. The effluent flows are assumed to be at thermodynamic equilibrium as is shown in Fig. A-3, although non-equilibrium approaches can be worked out [65].

Each stage can be described by a set of $2c+3$ MESH[3] equations, where *c* is the number of components. Sometimes, the set of equations can be reduced further, for instance by substituting the equilibrium relations in the species mass balances. For *n* stages, we have $n(2c+3)$ equations. For chromatographic systems, where typically $n = 100-500$ theoretical trays and $c = 3$ components (binary mixture in solvent), this leads to 900–4500 equations. These have to be solved simultaneously using a multivariate Newton method, with special matrix handling procedures to reduce the amount of stored data. Programs that can solve these equations are available commercially such as ASPENTECH and ChemSep. A special version of the latter program is available for SMB-simulation[4].

Fractionating Reactors

Models for fractionating reactors have the same structure as those of non-reactive systems. Reaction terms, however, are often highly non-linear and couple the

[3] $2c+3$ MESH equations: species and overall Mass balances $(c+1)$, Equilibrium relations (c), Sum-of-mole fraction relations (1) and, when applicable, an entHalpy balance (1). For further details, see Refs. [65, 66].

[4] Please contact one of the authors (LW) for updated details: L.A.M.vanderWielen@tnw.tudelft.nl.

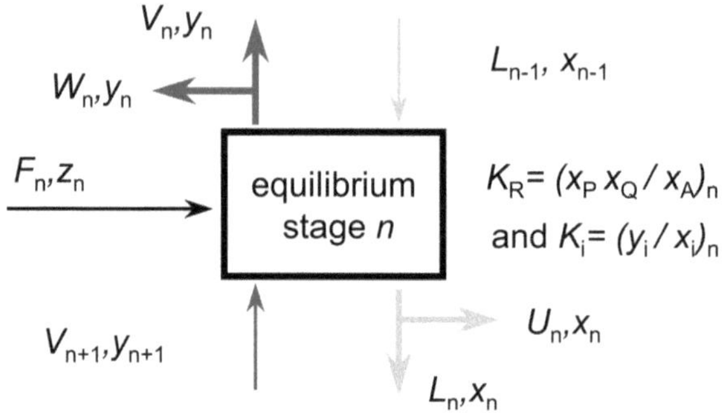

Fig. A-4. Schematic representation of a *reactive* equilibrium stage with a feed (F) and side withdrawal streams (U, W)

various equations more intimately. We use the same equilibrium stage model as discussed above, but supplement the reaction details as well (Fig. A-4).

The mass balance for species i at stage n without W and U streams, now reads:

$$V_{n+1} y_{i,\,n+1} + L_{n-1} x_{i,\,n+1} + F_n z_{i,\,n} = V_n y_{i,\,n} + L_n x_{i,\,n} + \nu_i R_n \tag{A a}$$

where νi is the stoichiometry coefficient of reacting species i and R is the reaction rate. νi is negative for reactants and positive for products. For equilibrium reaction (infinite rate), the reaction rate can be eliminated by adding mass balance equations pairwise for a substrate and a product. For the common equilibrium reaction of the type A = P + Q, this reduces the number of mass balances by one and extends the number of equilibrium relations by one. For constant distribution coefficients in a dilute system of species A, P and Q, we assume a mass action law-type phase equilibrium. The resulting set of equations reads as follows:

$$\begin{aligned} &V_{n+1}\,(y_{A,\,n+1} + y_{P,\,n+1}) + L_{n-1}\,(x_{A,\,n+1} + x_{P,\,n-1}) + F_n\,(z_{A,\,n} + z_{P,\,n}) \\ &\quad \cdot z_i = V_n\,(y_{A,\,n} + y_{P,\,n}) + L_n\,(x_{A,\,n} + x_{P,\,n}) \\ &V_{n+1}\,(y_{A,\,n+1} + y_{Q,\,n+1}) + L_{n-1}\,(x_{A,\,n+1} + x_{Q,\,n-1}) + F_n\,(z_{A,\,n} + z_{Q,\,n}) \\ &\quad \cdot z_i = V_n\,(y_{A,\,n} + y_{Q,\,n}) + L_n\,(x_{A,\,n} + x_{Q,\,n}) \\ &K_r\, x_{A,\,n} - x_{P,\,n}\, x_{Q,\,n} = 0 \end{aligned} \tag{A b}$$

where the reaction takes place primarily in the L-phase. These equations can be solved with the same procedure as outlined before. The initial estimate of the profile is now much more crucial. In some cases, it is required to use analytical solutions and special numerical techniques to solve the problem.

5
References

1. Datar R, Rosén C-G (1990) Downstream process economics. In: Asenjo JA (ed) Separation processes in biotechnology. Marcel Dekker, New York
2. Bruggink A (1998) Growth and efficiency in the (fine) chemical industry, Chimica Oggi 16:44–47
3. Straathof AJJ, Adlercreutz P (eds) (2000) Applied biocatalysis, 2nd edn. Harwood Academic, Reading, UK
4. Faber K (2000) Biotransformations in organic chemistry, 4th edn. Springer-Verlag, Berlin Heidelberg New York
5. Litjens MJJ, Straathof AJJ, Jongejan JA, Heijnen JJ (1999) Synthesis of primary amides by lipase-catalyzed amidation of carboxylic acids with ammonium salts in an organic solvent. Chem Commun 1255–1256
6. Diender MB, Straathof AJJ, van der Does T, Heijnen JJ (2001) Feasibility of a one-pot short-cut route from penicillin G to amoxicilin in anhydrous organic solvent (in press)
7. Wichmann R, Wandrey C, Bückmann AF, Kula MR (1981) Biotechnol Bioeng 23: 2789–2802
8. Seelbach K, Riebel B, Hummel W, Kula MR, Tishkov VI, Egorov AM, Wandrey C, Kragl U (1996) A novel efficient regeneration method of NADPH using a new formate dehydrogenase. Tetrahedron Lett 37:1377–1380
9. Schenkels P, de Vries S, Straathof AJJ (2001) Scope and limitation of the use of nicotinoprotein alcohol dehydrogenase for the coenzyme-free production of enantiopure fine-chemicals. Biocat Biotransform (in press)
10. Sybesma WFH, Straathof AJJ, Jongejan JA, Pronk JT, Heijnen JJ (1998) Reductions of 3-oxo esters by baker's yeast: current status. Biocat Biotransf 16:95–134
11. Stewart JD (2000) Organic transformations catalyzed by engineered yeast cells and related systems. Curr Opin Biotechnol 11:363–368
12. Freeman A, Lilly MD (1988) Effect of processing parameters on the feasibility and operational stability of immobilized viable microbial cells. Enzyme Microb Technol 23:335–345
13. van der Wielen LAM, Wesselingh JA (2002) Design booklet for separation processes (in press)
14. Den Hollander JL, Straathof AJJ, van der Wielen LAM (2001) Design of fractionating enzymatic reactors for hydrolysis reactions (in press)
15. Den Hollander JL, Aversente A, Diender MB, Straathof AJJ, van der Wielen LAM (2001) Enzymatic penicillin G hydrolysis in discretely contacted countercurrent water-butyl acetate biphasic systems (in press)
16. Den Hollander JL, Zomerdijk M, Straathof AJJ, van der Wielen LAM (2001) A continuous countercurrent enzyme reactor for the hydrolysis of penicillin G. (in press)
17. Keurentjes JTF, Voermans FJM (1997) In: Collins AN, Sheldrake GN, Crosby J (eds) Chirality in industry II: development in the manufacture and applications of optically active compounds. Wiley, Chichester
18. Overdevest PEM, van der Padt A, Keurentjes JTF, van't Riet K (1999). In: Scamehorn JF, Harwell JH (eds) Surfactant based separations: science and technology. ACS symposium series 740, ACS, Washington DC
19. Mazzotti M, Storti G, Morbidelli M (1997) Supercritical fluid simulated moving bed (SF-SMB) chromatography. J Chrom A 786:309
20. Jensen TB, Billiet HAH, van der Wielen LAM (1999) Method of substantially continuously separating two compounds using a (simulated) moving bed. UK Patent 2826684
21. Jensen TB, Reijns T, Billiet HAH, van der Wielen LAM (2000) A novel SMB method for reduced solvent consumption. J Chrom A 873:149–162
22. Houwing J, Jensen TB, van Hateren SH, Billiet HAH, van der Wielen LAM (2001) Salt gradients in SMB for protein separations. AIChE J (in press)

23. Lebreton B, Zomerdijk M, Ottens M, van der Wielen LAM (1999) Effect of impurities upon crystallisation kinetics of β-lactam antibiotics. Proc Annual Meeting AIChE, November 1999
24. Ottens M, Lebreton B, Zomerdijk M, Bruinsma D, van der Wielen LAM (2001) Crystallization kinetics of Ampicillin. Ind Eng Chem Res 40:821-4827
25. ■
26. Wesselingh JA, Krishna R (2000) Mass transfer in multi-component mixtures. Delft University Press
27. van Halsema FED, van der Wielen LAM, Luyben KCAM (1997) The modelling of carbon dioxide aided extraction of carboxylic acids from aqueous solutions. Ind Eng Chem Res 37:748-758
28. Greve A, Kula M-R (1991) Cost structure and estimation for the recycling of salt in a protein extraction process. Bioproc Eng 13:173-177
29. ■
30. Liu C-L, Kamei D-T, Wang JA, Wang DIC, Blankenschtein D (1998) Separation of proteins and viruses using aqueous two-phase micellar systems. J Chrom B 711:127-138
31. Presson J, Nystrom L, Ageland H, Tjerneld F (1999) Purification of recombinant proteins using thermoseparatig aqueous two-phase system and polymer recycling. J Chem Technol Biotechnol 74:28-243
32. van Berlo M, Ottens M, Luyben KCAM, van der Wielen LAM (2000) Feasible boundaries of aqueous two-phase systems with NH_3 and CO_2 as recyclable volatile salts. Biot Bioeng 70:65-71
33. de Vroom E, Kers EE, Heijnen JJ (1998) Method for separation of solid compounds in suspension. European Patent EP0997199A1
34. Jauregi P, Hoeben M, van der Lans RGJM, Kwant G, van der Wielen LAM (2000) Selective interfacial fractionation of near identical crystal mixtures (in press)
35. Jauregi P, van der Lans RGJM, Hoeben M, van der Wielen LAM (2000) Selective interfacial fractionation of near identical particle mixtures. UK Patent Application
36. Braas GMF, Walker SG, Lyddiatt A (2000) Recovery in aqueous two-phase systems of nanoparticles applied as surrogate mimics for viral gene therapy vectors. J Chrom B 743:409-420
37. Oyama K (1987) Enzymatic synthesis of aspartame in organic synthesis. In: Laane C, Tramper J, Lilly MD (eds) Biocatalysis in organic media. Elsevier, Amsterdam, pp 209-224
38. ■
39. Johansson HO, Karlström G, Tjerneld F, Haynes CA (1999) Driving forces for phase separation and partitioning in aqueous two-phase systems. J Chrom B 711:3
40. van der Wielen LAM, Rudolph SJG (1999) On the generalization of thermodynamic properties for selection of bioseparation processes. J Chem Biochem Technol 74:275-283
41. van Buel MJ, Gude M, van der Wielen LAM, Luyben KCAM (1997) Gradient elution in CPC. J Chrom 773:13-22
42. Jansen ML, Hofland GW, Houwers J, Straathof AJJ, van der Wielen LAM, Luyben KCAM, van den Tweel WJJ (1996) Effect of pH and concentration on column dynamics of weak electrolyte ion exchange processes. AIChE J 42:1925-1937
43. van der Wielen LAM, Diepen PJ, Houwers J, Luyben KCAM (1996) A countercurrent adsorptive reactor for acidifying bioconversions. Chem Eng Sci 51:2315-2325
44. Paiva AL, Malcata FX (1997) Integration of reaction and separation with lipases - an overview. J Mol Cat B-Enzym 3:99-109
45. Mazzotti M, Kruglov A, Neri B, Gelosa D, Morbidelli M (1996) A continuous chromatographic reactor: SMBr Chem Eng Sci 51:1827-1836
46. Kawase M, Suzuki TB, Inoue K, Yoshimoto K, Hashimoto K (1996) Increased esterification conversion by application of the simulated moving-bed reactor. Chem Eng Sci 51:2971-2976
47. Mensah P, Carta G (1999) Adsorptive control of water in esterification with immobilized enzymes. Continuous operation in a periodic counter-current reactor. Biotechnol Bioeng 66:137-146

48. Gude MT, Meuwisen HHJ, van der Wielen LAM, Luyben KCAM (1996) Partition coefficients and solubilities of α-amino acids in aqueous 1-butanol solutions. Ind Chem Eng Res 35:4700–4712
49. ■
50. Prokopakis GJ, Asenjo JA (1990) Synthesis of downstream processes. In: Asenjo JA (ed) Separation processes in biotechnology. Marcel Dekker, New York
51. ■
52. ■
53. Sheldon RA (1993) Chirotechnology. Marcel Dekker, New York
54. ■
55. Bruggink A (1996) Biocatalysis and process integration in the synthesis of semi-synthetic antibiotics. Chimia 50:431–432
56. Bruggink A (2000), Green solutions for chemical challenges; biocatalysis in the synthesis of semi-synthetic antibiotics. In: Zwanenburg B, Mikolajczyk M, Kielbasinsky P (eds) Enzymes in action. NATO sciences series 1/33, Kluwer Academic, pp 449–458
57. Bruggink A, Roos EC, de Vroom E (1998) Penicillin acylase in the industrial production of β-lactam antibiotics. Org Process Res Dev 2:128–133
58. Diender MB, Straathof AJJ, van der Wielen LAM, Ras C, Heijnen JJ (1998) Feasibility of the thermodynamically controlled synthesis of amoxicillin. J Mol Catalysis B: Enzymatic 5:249–253
59. Schroën CGPH, Nierstrasz VA, Kroon PJ, Bosma R, Janssen AEM, Beeftink HH, Tramper J (1999) Thermodynamically controlled synthesis of β-lactam antibiotics. Enzyme Microb Technol 24:498–506
60. Nierstrasz VA, Schroën CGPH, Bosma R, Kroon PJ, Beeftink HH, Janssen AEM, Tramper J (1999) Thermodynamically controlled synthesis of Cefamandole, biocatalysis and biotransformation 17:209–223
61. Diender MB, Straathof AJJ, van der Does T, Zomerdijk M, Heijnen JJ (2000) Course of pH during the formation of amoxicillin by a suspension-to-suspension reaction. Enzyme Microb Technol 27:576–582
62. Kemperman GJ, de Gelder R, Dommerholt FJ, Raemakers-Franken PC, Klunder AJH, Zwanenburg B (1999) Clathrate type complexation of cephalosporins with β-naphthol. Chemistry 2163–2168
63. Kemperman GJ, de Gelder R, Dommerholt FJ, Raemakers-Franken PC, Klunder AJH, Zwanenburg B (1999) Design of inclusion compounds of cephalosporin antibiotics (in press)
64. Storti G, Mazzott M, Morbidelli M, Carra S (1993) Robust design of binary countercurrent adsorption processes. AIChE J 39:471–492
65. Taylor R, Krishna R (1993) Multi-component mass transfer. Wiley, New York
66. Seader JD, Henley EJ (1998) Separation process principles. Wiley, New York

Received: February 2002

Note Added in Proof

Unfortunately this review article was not proofread. Despite many requests from us to the main author we never received the imprimatur from him, in particular, he never sent the missing references. We sincerely apologize for this incomplete contribution but did not wish to wait any longer and allow the other contributions to become more and more out of date. We hope you will understand our position.

Dr. Marion Hertel
Senior Editor Chemistry, Springer-Verlag

Membrane-Assisted Extractive Bioconversions

Pedro Fernandes[1,2] · Duarte M.F. Prazeres[1] · Joaquim M.S. Cabral[1]

[1] Center for Biological and Chemical Engineering, Instituto Superior Técnico, Av. Rovisco Pais, 1049-001 Lisboa, Portugal. *E-mail: joaquim.cabral@ist.utl.pt*

[2] Universidade Lusófona de Humanidades e Tecnologias, Av. do Campo Grande 376, 1749-024 Lisboa, Portugal

This chapter summarizes the use of membrane reactors in extractive bioconversions as process integration systems leading to in situ product recovery. Several membrane reactor configurations are analyzed, taking into account the type of bioconversion, biocatalyst type and location (either in the aqueous phase or in the membrane), membrane chemistry and morphology, solvent (extractant) type and its biocompatibility. Modeling of liquid-liquid extractive membrane bioreactors operation is also analyzed considering kinetics and mass-transfer aspects. The chapter includes examples from the authors' laboratory as well as other published in the field. Both enzyme and whole cell-based bioconversions are considered. Relevant aspects related to the solvent (extractant) toxicity and how the membrane could protect the biocatalytic activity are analyzed. Trends in this field are also given.

Keywords. Extractive bioconversions, Membrane reactors, Bioreactors, Process integration

Advances in Biochemical Engineering/
Biotechnology, Vol. 80
Series Editor: T. Scheper

Abbreviations

6-APA	6-aminopenicilannic acid
AOT	sodium di(2-ethylhexyl) sulfosuccinate
EO	ethylene oxide
HF	hollow fiber
HPS	hydroxypropyl starch polymer
MF	microfiltration
MWCO	molecular weight cut-off
PBHF	packed bed hollow fiber
PEG	polyethylene glycol
PO	propylene oxide
UF	ultrafiltration

1 Extractive Bioconversions

1.1 Introduction

The use of microorganisms as useful product manufacturing tools has been known to man since remote ages [1]. However, only in the nineteenth century were the scientific foundations of fermentation processes laid. The first useful compound other than ethanol to be produced in industrial scale fermentations was lactic acid, shortly followed by acetone, butanol, citric acid, and gluconic acid, in a time period ranging from late 1880 up to 1940 [2]. Nevertheless, the emerging petrochemical industry made the known fermentation processes uneconomical [2]. Only in the early 1940s did the technological development of microbial-assisted transformations take off, mainly due to the large need for penicillin [3]. Another major step followed shortly after, in the 1950s, when microbial transformation of steroids reached industrial scale production [4]. These milestones set a path leading to the replacement of some classical chemical processes by microbial-mediated transformations. Besides the use of the multireaction sequences of fermentation processes in which a product is formed de novo from substrates such as molasses or monosaccharides, several useful bioconversions processes were performed in which a compound is converted into a structurally related product by one or a small number of cell-contained enzymes or by enzyme preparations [1,5]. Traditionally, fermentations and bioconversions were performed in aqueous media, which severely limited the application of microbial-based processes to industrial scale processes, due to the poor solubility of compounds of commercial interest. The combined use of non-conventional media coupled to membrane contactors provided a way to overcome such limitation, as will be thoroughly discussed in the following sections. These will focus on the use of synthetic membranes; thus, no particular reference will be made to biological membranes, since the discussion of these structures is beyond the scope of this work.

1.2
Bioconversion Limitations

Bioconversions present potential advantages such as enantio- and stereospecificity, reduced number of synthesis steps, mild reaction conditions, and less environmentally damaging waste products, when compared to the chemical approach [5–7]. However, some potential disadvantages are inherent to conventional bioconversions in aqueous media, namely low volumetric productivity, substrate and/or product inhibition or toxicity, and low solubility in aqueous media of organic compounds of commercial interest [6]. The bioreactor exiting stream thus presents a low product concentration, leading to increased complexity, and hence increased costs in the downstream processing. Some of these drawbacks, namely end-product inhibition and low volumetric productivities, made some fermentations of commercial interest, such as ethanol [1], butanol-acetone [8], or acetone-butanol-ethanol [9], particularly sensitive to competing chemical processes [1]. A successful approach to overcome these drawbacks, first presented in the late 1970s/early 1980s [10–14], was based on the continuous removal of the inhibitory end-products as these were formed. This goal was achieved using several approaches, namely:

- Evaporation of volatile fermentation products, by creating a vacuum in the fermenter, or applying vacuum to the broth in a separate vessel (flash fermentation), by pervaporation, which combines evaporation and permeation through a semipermeable membrane, or by stripping the toxic compounds from the fermentation broth into a gas directly sparged through the vessel.
- Product immobilization, by adsorption or specific binding into water-insoluble carriers or reversible complex formation with cyclodextrins, leading to insoluble compounds.
- Size selective permeation with membranes, allowing for high cell/enzyme density inside the reactor.
- Extraction into another phase, using water-immiscible organic solvents, a second aqueous phase, or supercritical fluids. The last of these techniques, with supercritical CO_2 as extractant, did not prove feasible on the extractive fermentation of 2-phenylethyl alcohol; cell harvesting was required prior to extraction [15]. No further data on the use of supercritical extractive bioconversions using whole cells was found.

To fully exploit the concept of in situ product recovery (ISPR), some of these methods were combined. Some examples of in situ product recovery techniques are referred to in Table 1. Application of ISPR in the production of organic acids,

Table 1. Some applications of in situ product recovery techniques

Technique	Product removed	Reference
Evaporation		
Vacuum fermentation	Ethanol	[17]
Flash fermentation	Ethanol	[18]
	Ethanol	[19]
Pervaporation	Ethanol	[20]
	Acetone-butanol-ethanol	[21]
Gas stripping	Ethanol	[22]
	Ethanol	[23]
	Acetone-butanol-ethanol	[24]
Product immobilization		
HP-20	Red pigment	[25]
XAD-4	3-Phenylcatechol	[26]
Affigel 601	L-erythrulose	[27]
Activated carbon	Fluorocatechol	[28]
XAD-7	Methylene dioxyphenyl isopropanol	[29]
Anion-exchange resins	Lactic acid	[30]
XAD-7	Androstadienedione	[31]
XAD-7, XAD-16	Solavetivone	[32]
Cyclodextrins	Butyric acid-acetic acid	[33]
XAD-4 and XAD-7	Ethanol	[14]
Size selective permeation		
Ceramic membrane	Lactic acid	[34]
	Ethanol	[35]
Mineral UF, 500 kDa cut-off	Propionic acid	[36]
Polysulfone, 10 kDa cut-off	6-APA	[37]
UF	Oligosaccharides	[38]
UF, 3 kDA	Sialyllactose	[39]
Phase extraction		
Aqueous-aqueous		
Extractant: PEG-phosphate	Xylanase	[40]
Extractant: (EO/PO)/HPS	Lactic acid	[41]
Extractant: PEG/dextran	Heat shock proteins	[42]
Extractant: PEG/dextran	Chitinase	[43]
Organic-aqueous		
Organic phase		
Hexane	4-Vinylguaiacol	[44]
Octanol, 30% (w/w) tridodecylamine	Citric acid	[45]
Isooctane	Phenylacetaldehyde	[46]
Hexadecane	1-Octanol	[47]
Oleyl alcohol, 20% (w/w) Hostarex	Butyric acid	[48]
Palm oil	Acetone-butanol-ethanol	[49]
Oleyl alcohol, 40% (w/w) trilauryl-amine (Alamine® 304 – 1)	Propionic acid	[50]
n-Octane	Styrene epoxydes	[51]
Oleyl alcohol	Ethanol	[52]
Hexadecane	Thiophene	[53]
Oleyl alcohol/Alamine 336	Lactic acid	[54]
AOT/Isooctane	Chymotrypsin	[55]
Oleyl alcohol	Ethanol	[56]
Supercritical		
Extractant: CO_2	Ethyl myristate	[57]

low chain alcohols, cellulases, and monoclonal antibodies led to improvements in yield and productivity of 1.4–6 relative to conventional processes [16].

High reaction rates could thus be maintained, and more concentrated feeds were allowed for fermentation [58, 59]. The integration of part of the downstream processing in the bioreactor, allowing for ISPR-decreased product recovery costs [58], thus increased the competitiveness of the biochemical process.

1.3 Liquid-Liquid Extractive Bioconversions

The effectiveness of liquid-liquid extractive biocatalysis was confirmed during the 1980s with a great deal of work directed towards solvent selection [60]. Product recovery from the fermentation media using liquid-liquid extraction was already an established unit operation on the downstream processing of fermentation products [3]. The desirable organic solvent characteristics focused basically on its physicochemical properties. However, the use of organic solvents in an in situ recovery process also required low toxicity towards the biocatalyst (Table 2).

Much work has been done recently to understand the toxic effects of organic solvents on biocatalysts, their response mechanisms, and the development of adequate parameters for the classification of organic solvents in terms of biocompatibility, based on some of their physicochemical characteristics. This has been the subject of recent extensive reviews [62–70], and a detailed analysis of these factors is beyond the scope of this work. However some fundamentals should be referred to. Thus, it is generally accepted that the main target for solvent toxic effect in whole cells lies in the biological membrane. Solvents tend to accumulate there, disturbing their integrity and ultimately their physiological function. As for enzymes, the inactivation effect of the solvents is related to their ability to strip the essential water of enzyme molecules.

Despite the general toxic effect of solvents to whole cells, some microbial strains present an unusual tolerance to solvents (e.g., *Pseudomonas* [69, 71], *Rhodococcus* [72]). This may involve an adaptation mechanism at the level of the cytoplasmatic membrane, aiming to restore its stability and fluidity, once dis-

Table 2. Aimed solvent characteristics for use in extractive bioconversions (adapted from [64, 65])

Biocompatibility
Favorable partition coefficient for the product
High selectivity
Low emulsion-forming tendency
Low aqueous solubility
Non-environmentally hazardous
Non-toxic for humans
Available in bulk quantities at low cost
Physical-chemical stability
Allowing for easy product recovery
Non-biodegradable

turbed by the solvent. Also, changes in the lipopolysaccharide content of the outer membrane, or the development of mutant strains with different porines compared to native strains, when Gram-negative cells are concerned, were reported [70, 73]. These mechanisms may be considered passive, but a solvent efflux pump was described in *Pseudomonas* strains [74, 75], which actively excreted solvents. Finally, some strains were shown to be able to metabolize some organic solvents into a non-toxic or less toxic product [76–78].

Several attempts have been made to correlate the physical-chemical properties of organic solvents with their toxicity [62, 64, 67]. So far, the Hansch parameter ($\log P_{oct}$) has provided the more reasonable correlations with biocatalytic activity [67] and cell growth [76]. This parameter corresponds to the logarithm of the partition coefficient of the solvent in the water-octanol two-liquid phase system. Solvents are considered toxic if their $\log P_{oct}$ is below 2–3 (polar solvents), and non-toxic if their $\log P_{oct}$ is above 4–5 (apolar solvents). This correlation does not provide, however, an absolute rule and published data suggest that the toxicity of organic solvents may also be related to their molecular structure [79]. Furthermore, the toxicity depends on the microorganism used [79–81]. $\text{Log}\,P_{oct}$ values can be determined experimentally or calculated by Rekker's hydrophobic fragmental constant approach [82]. An on-line $\log P_{oct}$ calculation is also currently available at a website from the Syracuse Research Corporation (http://esc.syrres.com/interkow/kowdemo.htm).

Another key requirement for solvent selection for extractive bioconversions is the partition coefficient for the product (K_p) defined as the ratio of product concentration in the solvent to the product concentration in the aqueous phase at equilibrium. The higher the K_p, the higher is the product recovery capacity of the solvent. The use of lipophylic solvents as a second inert phase for the in situ product recovery of hydrophobic compounds, such as steroids [83], styrene epoxides [51], thiophene [53], or *n*-octanol [47], allowed both high extraction yields and biocatalytic rates, due to their biocompatibility. However, the extraction into these biocompatible solvents of water-soluble products, usually bulk chemicals of low specific cost such as ethanol and especially small chain length organic acids, is relatively ineffective due to the polarity of these compounds. Several approaches have been used to overcome this drawback. The yield of the extractive fermentation of ethanol using oleic acid as organic phase was increased by adding an enzymatic reaction system [84]. This promoted the esterification of ethanol with oleic acid, the higher hydrophobic character of the ester favoring extraction. The ethanol depletion resulting from the enzymatic reaction further shifted the thermodynamic equilibrium. The immobilization of the enzyme enhanced the overall yield [18]. The continuous extraction of ethanol using dodecanol was also performed, but, due to the low partition coefficient (0.35), a high organic to substrate feed phase ratio ($\approx$20) was required [85, 86].

ISPR in ABE (acetone-butanol-ethanol) fermentations have been carried out at laboratory scale using oleyl alcohol [87, 88] and decanol [89] as extractive agents, successfully enhancing solvent production by reducing end-product inhibition by butanol [90]. However, the high market value of the extraction solvents and the costs of solvent recovery prevented scaling-up. The use of methylated fatty esters (MFA) as extractive agents provided a good alternative [91].

Although the partition coefficients of ABE fermentation products was smaller in MFA (1.4 to 4-fold smaller) as compared to oleyl alcohol [92], cheap sources of esters, such as palm oil, were found. Furthermore, effluents from the palm oil industry provided adequate substrates for ABE fermentations [93], and since MFA could be employed as biodiesels, coupling ABE production with MFA to yield biodiesels, thus avoiding the costly solvent recovery step, was suggested [92].

A current approach for the recovery of organic acids involves the use of chemical extractive solvents. Among these, the more efficient may be divided in two groups: phosphorus-bonded oxygen donor extractants, such as alkyl phosphates and alkyl phosphine oxides, and aliphatic amine extractants [67, 50]. The extractive action of organophosphorous compounds is based on the solvation of the acid by donor bonds. Tributylphosphate (TBP) and trioctylphosphine oxide (TOPO) have been used for the extraction of organic acids, the latter proving more effective, which has been related to the presence of direct C–P linkages [94].

Aliphatic amine extractants react with organic acids forming ammonium salts or ion pairs, which are soluble in the organic phase. Among different amines, extractive fermentations with ternary amines, such as Alamine 336, are often performed, since these present low aqueous solubility and intermediate basicity. Such an approach provided a combined adequate extractive capacity for organic acids with the possibility of stripping [50]. The use of quaternary amines, such as Aliquat 336, as extractive agents is also widespread. Work performed on the extractive fermentation of lactic acid showed that these amines simultaneously extract both dissociated and undissociated forms of the organic acid [95]. However, the regeneration by back extraction proved difficult [50].

Although effective extractants, both aliphatic amines and organophosphorous compounds proved toxic to microorganisms. To reduce their toxic action, these extractants are blended, in adequate proportion, with low toxicity organic solvents, thus allowing for their use in extractive fermentations. The effect of diluents on the distribution coefficient was different according to the extractant used, as observed by Choudhury and co-workers [96], while evaluating the extractive efficiency of Aliquat 336 and trioctyl amine (TOA) in lactic acid extractive fermentation.

Solvent regeneration was performed by distillation [97], or, if either the solvent was essentially non-volatile (as with organophosphorous compounds or aliphatic amines) or the product was heat sensitive, solvent regeneration was carried out by back-extraction into water (hot water, preferentially) or alkaline solutions, if organic acids were to be recovered [98–100], or by adsorption to a solid phase [101, 102].

The advantages presented by water-immiscible organic solvents have extended from extractive fermentations/bioconversions to bioconversions of hydrophobic substrates in biphasic systems. The biocompatible organic solvent acts both as a substrate carrier and product extractant, thus allowing for high substrate concentrations and therefore increasing volumetric productivity. Both enzymatic and whole cell systems have been used for this goal, the latter being advantageous when multi-enzymatic pathways or co-factor regeneration are required [61].

Membrane reactors, in which the biocatalyst is physically retained behind a barrier, are often coupled to two-liquid extractive systems. This approach reduced the toxic effect of organic solvents, by avoiding phase toxicity effects and prevented emulsion formation, which often occurs in organic-aqueous two-liquid phase systems [16]. It further eased downstream processing, increased biocatalyst concentration, therefore volumetric productivity, enhanced biocatalyst stability, and allowed for its reuse or for continuous operation.

2 Membrane Bioreactors

2.1 Introduction

In membrane separation processes, a feed composed of two or more components is separated using a semi-permeable barrier, the membrane, into a permeate (the fraction of the feed that passes through the membrane) and a retentate (the part of the feed retained by the membrane). A membrane can thus be broadly defined as a selective barrier between two phases. This barrier can be made of a solid material or a fluid (gas or liquid). The use of membranes is common in classic bioconversion/fermentation processes, either in upstream processing – medium (including gas) filtration prior to entering the bioreactor – or in downstream processing, where microfiltration has been used for cell harvesting and cell debris removal [103]. Membrane processes are also commonly used for primary isolation [104], often coupled to solvent extraction, and in purification steps [105]. The integration of membranes in the bioreactor thus provided a logical attempt to gather in a single operation bioconversion, product recovery, and/or concentration and biocatalyst recovery, a goal that has been successfully achieved and has found a wide range of applications.

Membrane bioreactors were developed around the concept of physically separating biocatalyst and substrates and/or products using a semi-permeable synthetic membrane. The biocatalyst is thus confined to a defined zone in the membrane reactor, while substrates and products flow across the membrane either by diffusion (induced by concentration gradients) or by convection (generally induced by pressure gradients). These characteristics led to the early use of membrane reactors for enzymatic hydrolysis of macromolecules, such as starch or cellulose [106–108]. This trend has been maintained and recently Mountzouris and co-workers evaluated the use of an ultrafiltration stirred cell module as a mean to control product molecular sizes and characteristics derived from the enzymatic depolymerization of dextran [109]. Further developments were performed in bioconversions in non-conventional media. The combined use of organic-aqueous two-liquid phase systems and membrane modules allows for an interfacial contact area, the membrane acting as an interfacial catalyst, while physically separating the two phases [110]. Cells are easily contained behind the membrane as a result of an adequate choice of the membrane pore size. Enzymes have been directly immobilized onto the membrane by physical adsorption [111–114], or confined to one side of the membrane by enlargement through immobilization in an intermediate support (reversed micelles [115–117], adsorption to particulate material [118, 119]), isoelectric focusing [120] or size exclusion [121]).

2.2 Classification of Membrane Reactors

Membrane reactors were classically grouped according to the hydrodynamics/configuration of the system in CSTR and PFR types [106]. However, this proved unable to comprise some commonly used types in UF, such as flat membranes or dead-end operated modules and multiphase bioreactors. A classification based on the contact mechanisms that bring together substrate and biocatalyst was thus proposed [110]. Thus, membrane reactors could be divided into direct contact, diffusion contact, and interfacial contact reactors.

In direct contact reactors, substrate and biocatalyst are on the same side of the membrane and therefore diffusional resistances can be avoided if free biocatalyst is used or external mass transfer resistances is minimized, if the biocatalyst is used in an immobilized form (Fig. 1). Among the more common of these reactor types is the recycle reactor, basically a membrane, usually a hollow-fiber or a tubular module, connected to a stirred vessel in a semi-closed loop configuration. The substrate and the enzyme (if in a free form) were continuously recycled from and to the reaction vessel, while the product permeated through the membrane unit; thus, the whole system was performing as a CSTR [110]. Also included in this group are dead-end and dialysis reactors. In the former, reaction and separation take place in the same compartment. The substrate solution was continuously fed under pressure to a cell unit containing the biocatalyst and a suitable membrane, through which the product permeates. Since a low membrane area to reactor volume is available and agitation of the bulk phase is required to reduce concentration-polarization phenomena, the use of this equipment has been restrained to the laboratory scale. In the dialysis reactor, two process streams flow in each side of the membrane: one for substrate feeding, the other for product removal, which permeates through the membrane by a solution-diffusion mechanism. Low mass transfer rates due to the diffusion-based mass transfer mechanism were the main drawback of this set-up. Among these membrane reactors, a particular sub-class can be considered involving a hybrid membrane-emulsion reactor [121, 122]. In these reactors, an integrated unit of organic-aqueous two-liquid phase reaction system and membrane separation module was used with the aim of improving bioreactor efficiency. Thus, substrates were fed into the organic phase, which was dispersed in the aqueous phase containing the biocatalyst, the substrates partitioning into the aqueous phase. The product formed was extracted back into the organic phase. The reactors were operated as ultrafiltration cells, the organic phase being separated once product extraction had taken place [121].

Diffusion contact reactors are limited to the bioconversion of low-molecular weight substrates. In these reactors, the biocatalyst is contained behind the membrane, thus substrate has to diffuse through this barrier in order for bioconversion to occur (Fig. 2). The product formed diffuses back to the unreacted substrate stream. The use of these reactors is further restricted since mass transfer of substrate is diffusion based, thus permeation of substrates through the membrane is often the overall rate-limiting step [123]. Diffusion membrane reactors have also been coupled to organic-aqueous two-liquid phase systems, the two

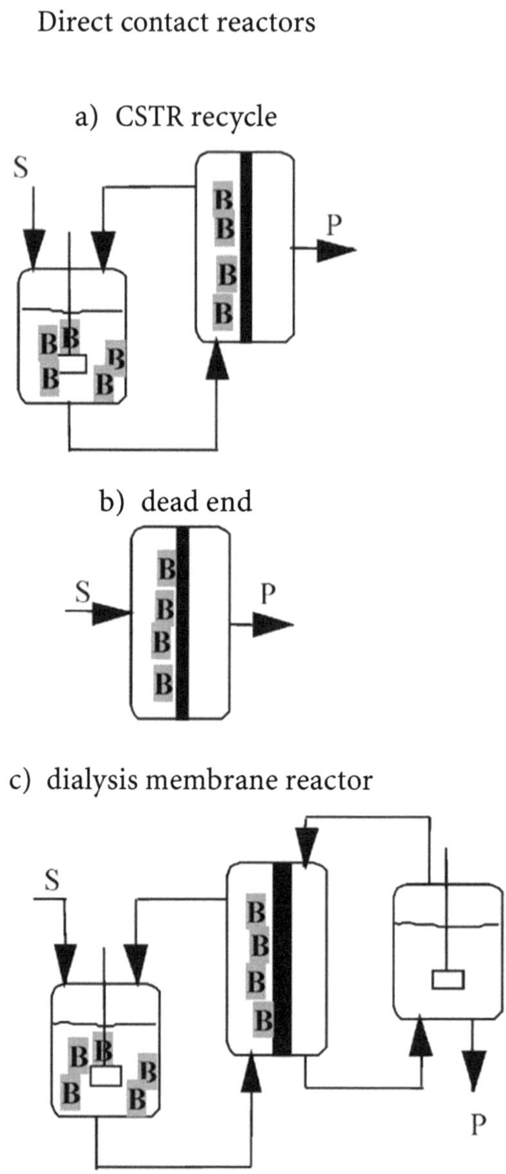

Fig. 1a–c. Direct contact membrane reactors (adapted from [110]). *B* biocatalyst, *S* substrate, *P* product

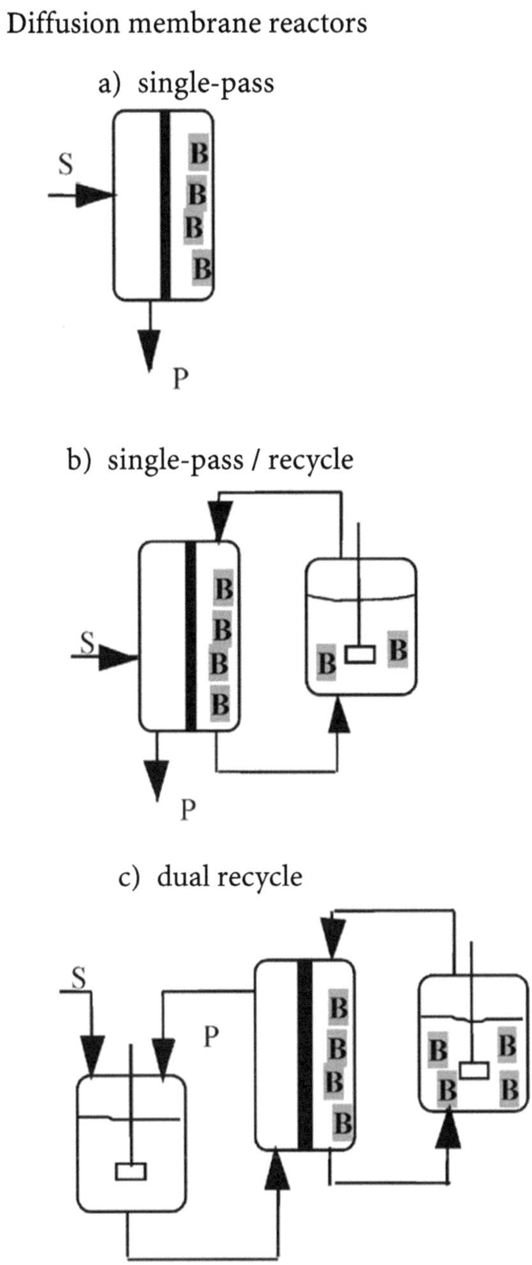

Fig. 2 a–c. Diffusion membrane reactors (adapted from [110]). *B* biocatalyst, *S* substrate, *P* product

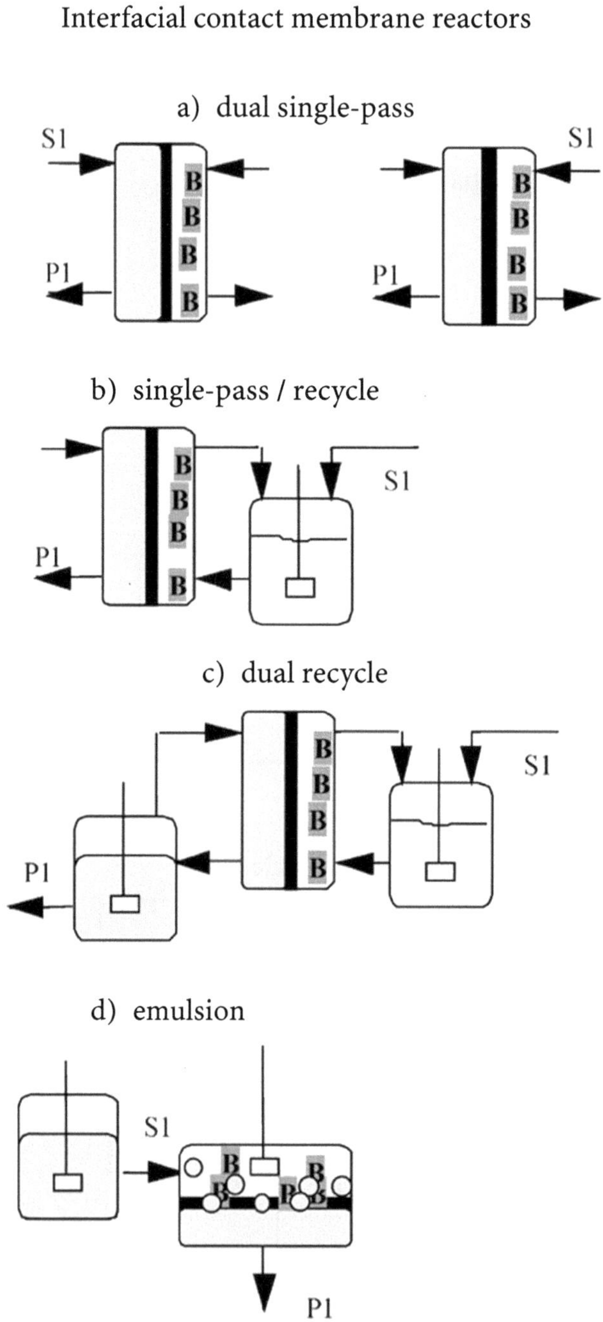

Fig. 3a–d. Interfacial contact membrane reactors (adapted from [110]). *B* biocatalyst, *S* substrate, *P* product. *Shaded area*: organic phase, *white area*: aqueous phase. Dead-end reactor is a hybrid system, combining an emulsion reactor and a membrane module in the same unit, through which the organic phase flows, while aqueous phase is rejected

liquid phases acted as a reservoir for substrates and/or products and the macroscopically non-porous dense membrane used to separate the two phases [124]. Biocatalyst-substrate contact is thus promoted by a solution-diffusion mechanism.

Diffusion contact reactors can be further subdivided according to the flow pattern of the enzyme- or substrate-containing stream. The enzyme-containing stream can be either confined or recirculated through the system, whereas the substrate-containing stream can either flow through the membrane module in a single pass or in a recirculation mode.

In interfacial contact reactors, a selectively wetted porous membrane is used to maintain an organic-aqueous interface in the plane of the membrane, while allowing for interfacial contact between the substrate and the biocatalyst (Fig. 3). Bulk mass transfer limitation, common in conventional heterogeneous emulsion systems, could thus be reduced [125]. Once more, the two liquid phases acted as a reservoir for substrates and/or products. To keep the interface in the plane of the membrane, a slight positive pressure in the non-wetting phase was needed [126, 127].

Organic and aqueous streams may flow through the membrane module in a single pass or be recirculated with external vessels. Alternatively, one of the streams may be recirculated, while the other either flows in a single pass or is kept in a batch mode.

Examples of these applications are listed in Table 3.

2.3 Membrane Chemistry and Morphology

Membrane classification can be done according to several viewpoints. A major division can be made between biological and synthetic membranes. Biological membranes are semi-permeable barriers that separate either the inside from the outside of the cell, or enclose internal cell structures, but these will not be addressed in this work. Commonly used membranes in separation or bioconversion processes are made of synthetic polymers or ceramics (Table 4).

Traditionally, membrane pore size was limited to the microfiltration range, corresponding to 0.1–10 μm [160], and ultrafiltration range, corresponding to a nominal molecular weight cut-off (NMWCO) of 500–300,000 Da [106, 110], allowing for the retention of whole cells and most enzymes. Technological developments allowed for the manufacture of nanofiltration membranes, which have a NMWCO in the range 200–1000 Da [161]. These membranes can be either positively or negatively charged [162]. Nanofiltration is therefore a separation process based on the difference in both charge and size of the solutes [163]. Nanofiltration membranes were used for the separation of amino acids and peptides [164], lactic acid [165], and NADP(H) retention in continuous enzymatic synthesis, where it allowed for a 3.4 increase in the total turn-over number, as compared to an UF membrane [163].

Shifts in the pH of solutions thus led to specific permeation characteristics of the solutes through charged membranes [149]. The use of negatively charged nanofiltration membranes allowed the development of a continuous process of

Table 3. Some applications of membrane reactors

Membrane unit/Classification	Bioconversion	Reference
Direct contact reactors		
Tubular ceramic UF and MF/recycle	Enzymatic hydrolysis of haemoglobin	[128]
Tubular ceramic UF/recycle	Enzymatic hydrolysis of starch	[129]
Plate UF membrane/recycle	Enzymatic production of sialyllactose from colominic acid	[39]
UF Polysulfone HF/recycle	Enzymatic synthesis of hexanal	[130]
UF and MF tubular ceramic/recycle	Enzymatic transfructosylation of sucrose	[131]
UF and MF tubular ceramic/recycle	Enzymatic inversion of sucrose	[132]
UF polysulfone HF/recycle	Enzymatic hydrolysis of penicillin G	[37]
UF flat acrylic/recycle	Microbial oxidation of naphthalene	[133]
UF tubular ceramic/recycle	Lactic acid production	[134]
UF hydrophilic polyaramide/recycle	Enzymatic enantioselective reduction of 2-octanone	[135]
UF flat membrane	Fungal decolourisation of a waste sludge	[136]
HF Romicon/dead-end	Technetium Tc (VII) reduction with *E. coli* cells	[137]
Polymeric nanofiltration/dead-end	Enzymatic production of mannitol and gluconic acid	[138]
Nanofiltration/dead-end	Enzymatic production of L-glutamate	[139]
UF stirred cells/dead-end	Enzymatic production of oligodextrans	[109]
Dialysis	Lipolysis of olive oil	[116]
Dialysis	Cell-free protein synthesis	[140]
UF polyethylene flat membrane/dead-end	Enzymatic synthesis of aspartame precursor	[121]
UF polysulfone/recycle	Enzymatic resolution of amino acids	[141]
Diffusion contact reactors		
Tubular dense membrane/recycle	Whole cell-mediated reduction of geraniol to citronello	[124]
UF HF/single-pass	Multi-enzymatic fructose-1,6-diphosphate production	[142]
Tubular silicone rubber membrane/recycle	Bacterial treatment of metal-containing wastewaters	[143]
Tubular silicone rubber membrane/recycle	Bacterial treatment of chlorinated-wastewaters	[144]
UF cellulose HF/single-pass	Cell-free luciferase synthesis	[145]

Silicone rubber HF and spiral/dual recycle	Biodegradation of toluene	[146]
Polymeric UF/dual recycle	Biodegradation of phenols	[147]
Tubular dense membrane/recycle	Whole cell-mediated epoxidation of 1,7-octadiene	[148]
Interfacial contact reactors		
MF polypropylene HF/single pass	Hydrolysis of menhaden oil	[112]
Plasmaphan HF/single-pass: recycle	Esterification of dodecanol and decanoic acid with in situ water activity control	[119]
Charged MF/single-pass: recycle	Separation of amino acid derivatives	[149]
MF flat/dual recycle	Whole cell hydrolysis of menthyl acetate	[150]
Dialysis/dual recycle	Enzymatic resolution of (*S*)-(+)-Naproxen	[102]
UF regenerated cellulose HF/dual recycle	Yeast resolution of 1,2-epoxyhexane	[151]
Polypropylene HF/dual recycle	Propionic acid production	[100]
Polyamide HF /dual recycle	Lipolysis of triglycerides	[152]
Nylon and cellulose esters flat sheet membranes	Enzymatic hydrolysis of babassu oil	[153]
UF, MF, nanofiltration, reverse osmosis flat membrane/dead-end	Enzymatic interesterification between triacylglycerols and polyunsaturated fatty acids	[154]
Microporous polypropylene HF/dual recycle	Whole cell production of isovaleraldehyde	[155]
UF HF/dual recycle	Enzymatic synthesis of aspartame precursor	[156]

Table 4. Commonly used materials in membrane manufacturing

Material	Wetting characteristics	Manufacturer (Commercial name)	Reference
Polymers			
Polypropylene	Hydrophobic	Akzo AG/Enka (Accurel®)	[119, 126]
		Hoechst Celanese Corp. (Celgard®)	[155]
Teflon [a]	Hydrophobic	Gelman Sciences (TF)	[126]
Silicone [b]	Hydrophobic	BDH	[148]
Polyethylene	Hydrophobic	Tonen Chemical	[121]
Nylon 66	Amphiphilic	Gelman Sciences (Nylaflo)	[157]
Polyamide	Hydrophilic	Forschungsinstitut Berghof	[152]
		Romicon	[158]
Polyacrylonitrile	Hydrophilic	Sepracor	[125]
Polyethersulfone	Hydrophilic	Gelman (Supor)	[126]
		Polymer Science	[159]
Polysulfone	Hydrophilic	Hoechst Celanese	[126]
		Polymer Science	[159]
Cellulose	Hydrophilic	Asahi Medical	[145]
Regenerated cellulose	Hydrophilic	COBE Nephross	[151]
Inorganic			
ZrO_2-TiO_2	Hydrophobic	Orelis SA	[117]

[a] Teflon: polytetrafluoroethylene. [b] Composed of polydimethylsiloxane and silica.

enzymatic production of xylitol from D-xilose. By combining an efficient NADH regeneration with a high (over 98%) retention of NAD(H) through the use of the charged membranes, productivities of 80 g xylitol L^{-1} day^{-1} were obtained over a time period of 150 h, as compared with productivities of 35 g xylitol L^{-1} day^{-1} obtained in conventional fermentative processes [166].

Synthetic polymers are made by polymerization of one monomer or by the copolymerization of two different monomers. A broad range of structures has been produced, from linear chain polymers, such as polyethylene, to cross-linked structures, such as butyl rubber [167]. Polymer membranes can have symmetric or asymmetric structures. The former, which is considered the less important type, can be porous or microporous [168]. Dense membranes, such as silicone rubber, are non-porous on a macroscopic scale [124]; therefore, permeating species must dissolve into the polymer and then diffuse through the membrane, making them highly selective. However, mass transfer rates were much lower than those observed in porous membrane, due to the dominating solution-diffusion mechanism [167].

Porous membranes contain interconnected homogeneous sized pores. Symmetric porous membranes thus may present either a high permeability or a high separation factor towards small molecules, but not both. This drawback was overcome with the development of asymmetric membranes, in which there is a continuous shift in the pore size in one direction. Several subtypes are manufactured [168].

Asymmetric membranes are much more common. These membranes are made of two layers, a thin (0.1 – 1.0 μm thick) permselective layer, supported by a thicker (around 30 μm) microporous layer, which presented virtually no resistance to mass transfer [160]. Since the manufacture of the permselective membrane is sometimes defective rendering the membrane useless for the separation of gaseous mixtures, the permselective membrane may be coated with a permeable polymer. Composite membranes, made by interfacial polymerization, are composed of a thin cross linked top-layer, supported once more by a microporous support.

New materials have been used in the manufacture of polymeric membranes, allowing for enhanced separation yield and specificity. Enantioselective recovery of tryptophan and phenylethyl alcohol was achieved using a nobornadiene-based membrane containing optically active groups [161]. The optical resolution of tryptophan enantiomers, through the use of polypyrrole membranes containing polymeric counterions, acting as molecular recognition agents for the separation of the enantiomers, was also recently reported [169]. Polymer-composite membranes have also been developed for the selective transport of cations or anions [170]. These are made of polyamine-terminated dendrimers grafted onto gold-coated silicone wafers. According to the pH, either -NH_2 groups are protonated and thus reject cations, or -COOH groups are deprotonated and thus reject anions.

Inorganic membranes, usually applied when high temperatures or chemically active mixtures are involved, are made of ceramics [171, 172], zirconia-coated graphite [173], silica-zirconia [174], zeolites [168], or porous glass [175] among others [176]. Ceramic membranes are steam sterilizable and offer a higher mechanical stability [134], thus they may be preferably used in aseptic fermentations, since some hollow fibers are only chemically sterilizable and not very suitable for reuse. Composite materials, in which glass fiber filters are used as support for the polymerization of acrylamide monomers, were developed for the hydrolysis of penicillin G in an electrically immobilized enzyme reactor. By careful adjustment of the isoelectric point of amphoteric membranes, the product of interest (6-aminopenicillanic acid) was retained in an adequate chamber, adjacent to the reaction chamber, while the main contaminant (phenyl acetic acid), was collected in a third chamber [120].

Porous membranes can be incorporated into compact modules with several shapes [167]. Hollow-fiber devices are, by far, the most widespread modules, mainly due to the high packing density ($500-9\times10^3$ m^2 m^{-3}) and low cost. They are, however, prone to fouling and are difficult to clean relative to other modules [167]. Plate and frame and tubular modules are also used [103].

2.4
Use of Membrane Reactors for Process Integration

As an immobilization method, both for whole cells or enzymes, membrane bioreactors provide the advantages and drawbacks common to entrapment or adsorption methods. They nevertheless present particular assets. Mass transfer in the porous supports generally used (alginate, k-carrageenan, zeolites, silica) is a diffusion-controlled process, often becoming the overall rate-limiting step. This may be overcome by the use of membrane modules. This equipment also avoids the need for complex, costly, and time-consuming sterilization and immobilization procedures. Furthermore, membrane entrapment is a mild immobilization technique, requiring neither chemical agents nor harsh environmental conditions, while it may also allow the entrapment of co-factors, required for carrying out co-enzyme-dependent bioconversions.

Membrane modules are however prone to fouling or concentration-polarization phenomena, which considerably decreases permeation flux and mass transfer [103, 177]. Furthermore, when polymeric membrane contactors are coupled to biocatalytic organic-aqueous two-liquid phase systems swelling of the membrane was observed, the extension of such phenomena depending on the nature of the solvent used [126]. This is an irreversible effect and therefore requires membrane replacement.

The common approach to maintain minimal polarization is to operate at high shear rates [103]; however, this can be harmful to biocatalysts. To decrease gel-layer formation in the surface of the membrane, Hakoda and co-workers [178] applied an electric field of 50 and 100 V to a ceramic membrane module used in the lipolysis of triolein in a reversed micellar system. These authors reported a slight increase in the filtration flux (about 15%), without deleterious effects on enzyme stability for an operation time length of 12 h. The electrokinetic phenomena leading to the observations occurred even in apolar media, since small amounts of water or surfactant were present in such media [178].

Fouling is caused by any species (organic, inorganic, or biological) that interacts with the membrane. It is often an unpredictable phenomenon, depending on the membrane nature (pore dimensions, structure, and hydrophilic/hydrophobic nature), operating times, media composition, and operational parameters (pH, temperature, transmembrane pressure). It was reported to cause up to 90% irreversible flux reduction in a few hours of operation; a rapid flux decrease is commonly observed once the operation is started [179]. The deposition of solid matter on the membrane cannot be rectified by changing operation parameters or membrane cleaning, and replacement is necessary to reverse the process of fouling [103, 175, 177, 179]. Membrane cleaning might be required daily, although some processes could be run for months before cleaning procedures were needed [179]. Cleaning procedures were required when a flow reduction of more than 10% or pressure losses over 15% were observed [180]. Furthermore, the larger transmembrane pressures observed in commercial systems, as compared to laboratory scale systems, made membrane cleaning in the former much more difficult [179]. Membrane cleaning can be performed by physical and/or chemical methods, in each case under harsh conditions. Physical cleaning is performed through the injection of water at the maximum allowable axial flow rate, and at the highest temperature and minimum pressure possible. Chemical cleaning is performed through the use of alkalis, acids, chelating agents, and disinfectants, depending on the nature of the precipitate. Usually, cleaning procedures combine several chemicals and are defined according to the nature of the membrane [180]. As an example, a chlorine solution was recirculated, at a temperature of 60 °C and a transmembrane pressure of 50 kPa for the cleaning of a ceramic fouled with a microbial polysaccharide [181]. Frequently cleaning the membrane has been reported to reduce membrane life irrespective of the method used [179, 180].

3 Liquid-Liquid Extractive Membrane Bioreactor Configurations

3.1 Selection of the Extraction Systems and Membrane Modules

In early biocatalytic systems developed for performing bioconversions in organic-aqueous two-liquid phase systems without emulsification, microporous polypropylene [182] or ultrafiltration-regenerated cellulose membranes [183] were used for lipase-catalyzed hydrolytic reactions. Although no comments were made concerning the criteria used for the selection of the membrane material, the need for a rationale underlying the careful choice of the membrane material in order for the membrane bioreactor to be successfully operated, was highlighted by Vaidya and co-workers shortly after [127, 157, 184], and later by Schroën and Woodley [126]. These authors focused on the membrane structure and wettability and how these related to the immobilization of the liquid-liquid interface within the membrane plane, which may require, as mentioned earlier, a slight positive pressure on the non-wetting liquid. An excessive pressure will lead, however, to the displacement of the wetting phase and ultimately to the breakthrough of the non-wetting liquid [127]. Thus, in order for an interfacial membrane reactor to perform in an effective and stable manner, a relatively high initial breakthrough pressure was required and its value should be constant, or otherwise increase throughout the course of the operation [127]. Schroën and Woodley [126] considered a breakthrough pressure in excess of 20–30 kPa high enough in order to be controlled during membrane separation of solvent/water mixtures. Giorno and co-workers carried out the enzymatic hydrolysis of triglycerides in an organic/aqueous asymmetric polyamide membrane reactor under a transmembrane pressure of 35 kPa [152]. Isono and co-workers carried out the synthesis of an aspartame precursor using a hollow-fiber membrane under a transmembrane pressure of 20 kPa in an organic/aqueous phase system [156].

The breakthrough pressure (P_b) through a pore of cylindrical cross-section can be related to the membrane pore size and to the interfacial tension at the liquid-liquid interface, according to the Laplace-Young law, Eq. (1) [127],

$$P_b = \frac{2\sigma_{wn}\cos\theta_{wm}}{r} \qquad (1)$$

where σ_{wn} is the interfacial tension at the liquid-liquid interface, θ_{wm} is the angle of contact between the wetting liquid and the membrane, and r is the pore radius at the line of contact.

The use of UF membranes is therefore advised, as compared to MF membranes, since the smaller pores of the former led to higher breakthrough pressures [157, 184]. However, microporous polypropylene membranes have been effectively used in several bioconversion systems [112, 113, 155, 184, 185], probably due to a high interfacial liquid-liquid tension in the reaction systems studied. A detailed theoretical discussion on the effects of pore geometry and the placement of the liquid phases in the breakthrough pressure can be found in an article by Vaidya et al. [127].

The presence of a biocatalyst, either whole cells [126] or enzymes [157], or any other biological surface-active materials either produced or present as substrates in the bioconversion system, such as fatty acids or long chain alcohols [127, 184], were expected to lower interfacial tension and hence breakthrough pressure [126, 157, 184]. A threefold decrease in the interfacial tension was observed in an aqueous-tetradecane system when either *Pseudomonas putida* or bakers' yeast cells were used, as compared to the cell-free system [126]. A decrease in the breakthrough pressure due to the presence of a surface-active agent, lauric acid, was also cited [184].

To have a clearer understanding of the effect of surface-active agents on the breakthrough pressure, it should be recalled that, besides a liquid-liquid interface, two other interfaces are formed: one between the wetting liquid and the membrane, the other between the non-wetting liquid and the membrane [127, 184]. Equation (2), yielding the effect of the two liquid-membrane interfaces and the breakthrough pressure was proposed by Vaidya and co-workers [127, 184],

$$P_b = \frac{2(\sigma_{nm} - \sigma_{wm})}{r} \tag{2}$$

where σ_{nm} is the interfacial tension at the non-wetting liquid-membrane interface and σ_{wm} is the interfacial tension at the wetting liquid-membrane interface.

The improvement of wettability of the membrane by the non-wetting liquid, ultimately leading to its breakthrough due to the adsorption of fatty acids at the interface between the non-wetting liquid and the membrane, was experimentally verified by Vaidya et al. [184].

The effect of the wetting characteristics of membrane bioreactors on the operation of organic-aqueous two-liquid phase systems was discussed by Vaidya and co-workers [127, 157, 184]. A similar discussion, but considering the use of membranes for separation of liquid/liquid mixtures downstream of a bioreactor, was carried out by Schroën and Woodley [126]. Some trends for the use of membranes in organic-aqueous two-liquid phase systems can be summarized from these works. The use of UF hydrophilic or amphiphilic membranes was usually advised for two-phase bioreactors [127], although fluoropolymer-based membranes could present an exception [126]. PTFE membranes, on the other hand, led to low breakthrough pressures, and therefore their use was limited.

Swelling of the membrane, resulting from its contact the organic solvent [186], also led to a decrease in the breakthrough pressure [126].

Porous membrane modules were therefore effectively used in bioreactors as an alternative to direct two-liquid contact systems, as long as phase breakthrough was avoided. This required a careful control of the transmembrane pressure, particularly if surface-active material was produced during bioconversions [126, 184, 187]. Fouling problems also developed in membrane-assisted multi-phase separation systems. This was observed by Conrad and Lee in the recovery of an aqueous bioconversion product from a broth containing 20% soybean oil by using ceramic membranes; fouling was caused mainly by soluble proteins and surfactants [188].

To minimize phase breakthrough, the use of dense membranes for bioconversion of hydrophobic molecules in organic-aqueous two-liquid phase systems was proposed by Doig and co-workers for the microbial reduction of geraniol [124] and the microbial epoxidation of octadiene [148]. These authors used a thin-walled (wall thickness = 250 μm [124] or 1 mm [148]) silicone rubber tubing, which separated the two liquid phases, while allowing small hydrophobic molecules to diffuse through the wall. Both fouling and bulk-phase breakthrough were avoided, and the membrane was impermeable to both macromolecular and ionic species. For this set-up to perform effectively, a high surface area was also required; thus, a coiled, small diameter tubing (2 mm [148] or 3 mm [124] inner diameter) was selected. A specific area of 35 $m^2\,m^{-3}$ of membrane was used [148], which is considerably lower than a specific area of 2.6×10^3 $m^2\,m^{-3}$ for a hollow-fiber module (estimated from data from Giorno and co workers [152]). The coiled tubing could be inserted in the bioreactor with the organic phase flowing in the lumen (tube) side [124] or otherwise immersed in a vessel containing the organic phase, with the biomass suspension flowing in the lumen side [148]. The latter flow pattern is preferred, since it generated a positive transmembrane pressure differential on the non-wetting phase, as long as this phase was recirculated at high Reynolds number [148]. Such experimental set-up allowed for high overall mass transfer coefficients [148], although biocatalyst deactivation could occur due to the high shear stress generated.

As for flow patterns, counter-current flow is preferred, since it allows a higher efficiency in mass transfer rate, therefore increasing the overall reaction rate. For instance, the initial hydrolysis rate of triglycerides catalyzed by a lipase in a membrane two-phase reactor was 60% higher when a counter-current mode was used, as compared to co-current mode [152].

3.2
Evaluation of Kinetics and Mass Transfer in Membrane Reactors

In membrane-based bioconversion systems, mass transfer resistance due to diffusion through the membrane and partition of the solutes over the membrane have to be added to the film resistances in the two liquid phases. An overall mass transfer coefficient (K) can thus be defined for each solute, based on the individual resistances. The concentration profile of a solute being extracted through a membrane is a function of some physical characteristics of the membrane, namely its wettability character, its geometry, porosity, and tortuosity, of the location of the aqueous and organic phases, and of the free diffusion coefficient of the solute, as evidenced in a detailed study performed by Prasad and Sirkar [189].

Assuming steady-state mass transfer, organic-aqueous liquid interface immobilized in the membrane pores, solute partition coefficient constant over the concentration range evaluated and immiscibility of the two liquids, global expressions for the overall mass transfer coefficients for hollow-fiber microporous membranes, based on the aqueous phase (K_w, Eq. (3)) or on the organic phase (K_o, Eq. (4)) can be derived [189].

$$\frac{1}{K_w} = \frac{d_1}{m_1\,d_2\,k_1} + \frac{d_3}{m_2\,d_{\ln}\,k_2} + \frac{d_4}{d_5\,k_3} \tag{3}$$

$$\frac{1}{K_o} = \frac{d_1}{d_2\,k_1} + \frac{m_1\,d_3}{d_{ln}\,k_2} + \frac{m_2\,d_4}{d_5\,k_3} \tag{4}$$

where d_{ln} stands for the logarithmic mean diameter of the hollow fiber, m_1 and m_2 are the partition coefficient of the solute, and k_2 is the membrane mass transfer coefficient for the solute that can be obtained by Eq. (5) [189].

$$k_2 = \frac{D\varepsilon}{L\tau} \tag{5}$$

where D stands for the diffusion coefficient in the membrane fluid phase (aqueous for hydrophilic membranes or organic for hydrophobic membranes) and ε, τ, L are the membrane porosity, tortuosity, and thickness, respectively. It is assumed that only unidimensional flow is present, and that the membrane is totally wetted by the selected phase.

The meanings of other terms are given in Table 5.

Table 5. Notation used for Eqs. (3) and (4) (adapted from [189])

	Membrane wettability	
	Hydrophobic	Hydrophilic
$1/K_w$		
Aqueous phase in lumen	$d_1 = d_3 = d_4 = d_5$ = hollow fiber inside diameter, d_2 = hollow fiber outside diameter, $m_1 = m_2$	$d_1 = d_2 = d_3 = d_4$ = hollow fiber outside diameter, d_5 = hollow fiber inside diameter, $m_2 = 1$
	k_1 = local mass transfer coefficient of the solute in organic phase on the shell side, k_3 = local mass transfer coefficient of the solute in aqueous phase on the tube side	
Organic phase in lumen	$d_1 = d_3 = d_4 = d_5$ = hollow fiber outside diameter, d_2 = hollow fiber inside diameter, $m_1 = m_2$	$d_1 = d_2 = d_3 = d_4$ = hollow fiber inside diameter, d_5 = hollow fiber outside diameter, $m_2 = 1$
	k_1 = local mass transfer coefficient of the solute in organic phase on the tube side, k_3 = local mass transfer coefficient of the solute in aqueous phase on the shell side	
$1/K_o$		
Aqueous phase in lumen	$d_1 = d_3 = d_4 = d_5$ = hollow fiber inside diameter, d_2 = hollow fiber outside diameter, $m_1 = 1$	$d_1 = d_2 = d_3 = d_4$ = hollow fiber outside diameter, d_5 = hollow fiber inside diameter, $m_1 = m_2$
	k_1 = local mass transfer coefficient of the solute in organic phase on the shell side, k_3 = local mass transfer coefficient of the solute in aqueous phase on the tube side	
Organic phase in lumen	$d_1 = d_3 = d_4 = d_5$ = hollow fiber outside diameter, d_2 = hollow fiber inside diameter, $m_1 = 1$	$d_1 = d_2 = d_3 = d_4$ = hollow fiber inside diameter, d_5 = hollow fiber outside diameter, $m_1 = m_2$
	k_1 = local mass transfer coefficient of the solute in organic phase on the tube side, k_3 = local mass transfer coefficient of the solute in aqueous phase on the shell side	

It should be pointed out that since very thinned-walled membranes were used, and also assuming a slight thickness of the liquid films, the contact areas were considered similar [190]. Overall mass transfer coefficients for flat membranes were thus easily derived from Eqs. (3) and (4), by setting all terms related to internal or external diameter equal to 1. Local mass transfer coefficients k_1 and k_3 were related only to the aqueous or organic phase.

To account for composite membranes, Prasad and Sirkar [189] presented more complex expressions to include the effect of both the hydrophobic and the hydrophilic moieties of these membranes on the overall mass transfer coefficients, according to Eq. (6) and Eq. (7).

$$\frac{1}{K_w} = \frac{1}{mk_1} + \frac{1}{mk_2} + \frac{1}{k_3} + \frac{1}{k_4} \tag{6}$$

$$\frac{1}{K_o} = \frac{1}{k_1} + \frac{1}{k_2} + \frac{m}{k_3} + \frac{m}{k_4} \tag{7}$$

where k_1 and k_3 are the local mass transfer coefficients in the organic and aqueous phases and k_2 and k_4 are membrane mass transfer coefficient for a solvent-filled pore or an aqueous-filled pore, respectively.

Local mass transfer coefficients in the shell side or in the tube side were determined using adequate correlations, some of which are presented in Table 6.

Table 6. Local mass transfer coefficients. D is the diffusivity, υ is the velocity, d is the hydraulic diameter, μ is the viscosity and ν is the kinematic viscosity and L is the length along the channel

Mass transfer correlations	Comments	Reference
Tube side (k_t)		
$k_t = 1.5(D^2 \upsilon d^{-2} L^{-1})^{1/3}$	Extraction of small solutes and proteins into a solvent. Valid for $8 < k_t d/D < 40$	[189]
$k_t = 1.64(D^2 \upsilon d^{-1} L^{-1})^{1/3}$	Absorption of gases into water. Valid for $3 < D^2 \upsilon d^{-1} L^{-1} < 500$	[189]
$k_t = 1.62(D^2 \upsilon d^{-1} L^{-1})^{1/3}$	Modification from Lévêque equation. Valid for $\rho \upsilon d \mu^{-1} \leq 2200$	[103]
$k_t = 2.906(D^2 \upsilon d^{-1} L^{-1})^{0.537}$	Transfer of trichloroethylene	[192]
$k_t = 0.02 D^{0.67} \upsilon^{0.88} d^{-0.12} \nu^{-0.55}$	Valid for $2500 < \rho \upsilon d \mu^{-1} < 10{,}000$	[103]
Shell side		
$k_s = a(1-\phi) D^{0.67} \upsilon^{0.66} L^{-1} \nu^{-0.33}$	Parallel flow on shell side in solvent extraction. For hydrophilic fibers, $a = 6.1$, for hydrophobic fibers $a = 5.85$. ϕ is the packing fraction of hollow fibers	[189]
$k_s = 8.8 D^{0.67} \upsilon d L^{-1} \nu^{-0.67}$	Parallel flow on shell side in solvent extraction for hydrophobic fibers	[193]
$k_s = 0.9 D^{0.67} \upsilon^{0.4} d^{-0.6} \nu^{-0.07}$	Cross flow, for $\phi \leq 0.07$	[194]
$k_s = 1.38 D^{0.67} \upsilon^{0.34} d^{-0.66} \nu^{-0.01}$	Cross flow, for $\phi \geq 0.07$	[192, 194]

Table 7. Solute mass balances for batch extraction through membranes in a two-liquid phase system (adapted from [99, 155, 189]). C_W and C_o stand for the bulk concentration in the aqueous and organic phases, respectively, A is the area for mass transfer, t is the time of operation, and m is a solute partition coefficient

	Overall mass flow	Solute mass balance to	
Extraction from		Organic phase	Aqueous phase
Aqueous phase	$N = K_o(m C_W - C_o) A$ $N = K_W(C_W - C_o m^{-1}) A$	$N = V_o \, dC_o/dt$	$N = - V_W \, dC_W/dt$
Organic phase	$N = K_o(C_o - m C_W)A$ $N = K_W(C_o m^{-1} - C_W)A$	$N = - V_o \, dC_o/dt$	$N = V_W \, dC_W/dt$

A detailed analysis on the development of correlations for mass transfer coefficients can be found in the literature [191].

Overall mass transfer coefficients were also calculated from experimental results [149, 155].

For batch extraction between the two liquid phases, and through the membrane, the time course of solute concentration can be calculated by solute mass balances to both aqueous and organic phases, assuming they are perfectly mixed. These are summarized in Table 7.

Combining the solute mass balances to the aqueous and organic phases and the overall mass flow equation, followed by integration, yielded an expression for the concentration profile [99, 150]. This allowed for the calculation of the overall mass transfer coefficient from the slope of a semi-log plot, as exemplified by Eq. (8) [155],

$$\ln\left[\frac{\left(1+\frac{V_W}{mV_o}\right)C_W - \frac{V_W}{mV_o}C_{W_0}}{C_{W_0}} = -\frac{AK_W}{V_W}\left(1+\frac{V_W}{mV_o}\right)t\right] \tag{8}$$

Where C_W and C_{W_0} stand for the solute concentration at time t and time 0, respectively, m is the solute partition coefficient, V_W and V_o stand for the volume of aqueous phase and organic phase, respectively.

3.3 Modeling

Mathematical models have been developed for predicting the performance of hollow-fiber membrane reactors [195–198]. Membrane bioreactor modeling was performed combining mass transfer and consumption/production terms [99, 155]. Integration of the resulting differential equation by using Runge-Kutta [116] or Euler [99, 155] methods led to simulated curves in good agreement with experimental results, involving the production of aldehydes [155] or organic acids [99] in hollow-fiber assisted two-liquid phase bioconversions, or the hydrolysis of olive oil in a dialysis system [116].

A mathematical model was developed by Wu and co-workers [125] to simulate the enzymatic resolution of a racemic mixture using a lipase entrapped in the sponge region of a hydrophilic hollow-fiber system. The material balances led to a set of linear ordinary differential equations, which was solved using a Runge-Kutta procedure, combined with a Newton-Raphson method, to generate theoretical concentration profiles of substrate and product effectively fitting the experimental data. A reaction-diffusion model was developed by Bouwer et al. [113] to predict product profiles in the hydrolysis of tricaprylin and peroxydation of caprylic acid, using a lipase immobilized in a membrane module. Mass balance modeling was also performed by Doig and co-workers [148] to optimize a bioreactor designed for the epoxidation of octadiene, using growing *P. oleovorans* cells in dense membrane bioreactors. Combining mass balances for substrates and products and assuming Monod kinetics, it was possible to predict biomass concentration and volumetric productivity.

4 Examples of Membrane-Assisted Extractive Bioconversions

4.1 Enzyme-Based Systems

Biocatalyst recovery and re-use or continuous operation, as well as integrated product recovery, are main assets for the economic viability of an industrial process. Much work has been carried out recently, mainly focusing on the transformation of oils and fats (Table 8).

The effect of the wetting characteristics of the membrane in the hydrolytic activity of immobilized lipase was evaluated by Bouwer and co-workers [113]. These authors suggested that the larger thickness of the reaction layer in hydrophilic membranes, as compared to hydrophobic membranes, allowed the retention of the full enzyme activity. A lipase adsorbed onto polypropylene membranes in a hollow-fiber system was used for the selective hydrolysis of menhaden oil [112]. The experimental set-up led to the release of about 88% of the fatty acid residues and retention of more than 90% of the aimed eicosapentaenoic and docosahexaenoic acids in a space-time of 3.5 h, while the half-life of the enzyme, under ideal operational conditions was 170 h. The enzymatic synthesis of an aspartame precursor in an organic-aqueous membrane-assisted two-liquid phase system has been continuously studied by Isono and co-workers [121, 149, 156, 201]. Productivity was increased from 6.6 to 8.4 kg m^{-3} d^{-1} [121].

Most of these bioconversions were carried out in conventional hollow-fiber modules, with the biocatalyst either immobilized in the membrane or freely suspended in the aqueous phase. However, the widespread use of hydrolytic enzymes for synthesis reactions [119, 202] presented a particular problem, since a shifting of the equilibrium was required to obtain a high yield of the synthesis product [119]. To effectively perform such bioconversions it was necessary both to remove water formed and to maintain water activity at the optimum level, which could be achieved by the use of salt hydrates [65]. On-line control of water activity was performed by continuously contacting the reaction medium with

Table 8. Examples of membrane-assisted enzymatic multiphase bioconversions

Membrane unit	Cut-off/pore size	Bioconversion system	Reference
Polyacrylonitrile/HF	50 kDa	Lipase-assisted production and crystallization of chiral compounds in biphasic system	[114]
Polyamide/HF	50 kDa	Lipase hydrolysis of olive oil in biphasic system	[152, 199]
Polypropylene/HF and plate and frame	100 nm	Lipase hydrolysis of triglycerides and peroxidation of fatty acids in biphasic system	[113]
HF		Antibiotic production from benzyl-penicillin using penicillin amidase	[200]
Polypropylene/HF	Microporous	Lipase hydrolysis of menhaden oil	[112]
HF	50 kDa	Synthesis of aspartame precursor in aqueous-organic extractive system	[156]
PBHF	500 nm	Lipase esterification of dodecanol and decanoic acid	[109]
Polyethylene/filtration cell	30–40 nm	Synthesis of aspartame precursor in aqueous-organic extractive system	[121]
Polypropylene and regenerated cellulose/PBHF		Lipolysis of lauric acid ethyl ester	[118]

the salt solutions. Wehtje and co-workers [203] developed an experimental set-up that proved the feasibility of this method, but it was considered unsuitable for large-scale operation [119]. A set-up based on a hollow-fiber module, in which the shell-side was filled with enzyme immobilized in microporous-sized support was proposed by Vaidya et al. [118], and effectively used for the esterification of dodecanol and decanoic acid in hexane, using polypropylene immobilized lipase, under continuous water activity control [119]. The saturated salt solutions, flowing through the lumen side, stripped the organic phase of the water formed during the reaction, and thus kept the water activity constant in the organic phase, and were again saturated by flowing through a packed bed of salt [119]. Recently, an experimental set-up was effectively used by Xin and co-workers to perform bioconversions in a low water activity medium [102]. These authors carried out the stereoselective hydrolysis of Naproxen methyl ester to produce Naproxen, using lipase immobilized in a modified porous silica support of low polarity. The immobilized biocatalyst was resuspended in the organic phase containing the substrate, and the product was transferred into an aqueous buffer flowing through a dialysis membrane tube placed inside the bioreactor.

4.2 Whole Cell Systems

Membrane-assisted extractive bioconversions have been mainly developed for systems leading to toxic end-products, mainly short chain alcohols and acids. In such systems, the organic phase acts as a product reservoir, removing it from the aqueous phase as it is produced. These systems were developed in order to comply basically with the needs of ethanol or organic acids fermentations. However, the increased use of whole cells as "bags" of enzymes to perform bioconversions, and the widening range of products resulting from these processes [204, 205] has expanded this concept.

The organic phase may also be used as a substrate reservoir, besides their use for product stripping from the aqueous phase. The effectiveness of membrane-assisted organic-aqueous two-phase bioconversions relative to direct-contact two-phase emulsion reactors was demonstrated by Westgate et al. [150]. These authors observed a fivefold increase in the maximum specific activity of hydrolysis of menthyl acetate catalyzed by *B. subtilis* cells when a 0.2 μm nylon flat membrane reactor was used, as compared to an emulsion reactor. This result was attributed to a continuous interfacial contact, which could only be achieved in an emulsion bioreactor at the cost of high power inputs. Doig and co-workers operated a dense membrane bioreactor for the production of citronellol from geraniol with a product accumulation rate similar to the one obtained in an emulsion reactor [124]. Some examples of membrane-assisted two-liquid phase bioconversions/fermentations are presented in Table 9.

A further development aimed at increasing process integration lies in the use of three-phase membrane reactors, which were developed mainly for the recovery of organic acids [208]. In these systems, the aqueous bioconversion medium,

Table 9. Examples of membrane assisted whole cell extractive bioconversions using organic solvents

Membrane unit	Pore size	Bioconversion system	Reference
Dense tubular silicone rubber membrane	–	Epoxidation of 1,7-octadiene with *Pseudomonas oleovorans* cells	[148]
Dense tubular silicone rubber membrane	–	Reduction of geraniol to citronellol with baker's yeast	[124]
Polypropylene/HF	Microporous	Oxidation of 2-methyl-1,3-propanediol	[99]
Polypropylene/HF	Microporous	Oxidation of isoamyl alcohol with *Gluconobacter oxydans* cells	[155]
Regenerated cellulose/HF	Microporous	Hydrolysis of 1,2-epoxyhexane with *Rhodotorula glutinis* cells	[151]
Polypropylene/HF	30 nm	Propionic acid production with *Propionibacterium acidipropionici* cells	[100]
Polypropylene/HF	30 nm	Biodegradation of trichloro-ethylene with *Methylosinus trichosporium* cells	[206]
HF	100 nm	Lactic acid production with *Lactobacillus delbrueckii* cells	[207]
Nylon/Flat membrane	200 nm	Hydrolysis of menthyl acetate with *Bacillus subtilis* cells	[150]

the organic extractive phase, and a second aqueous stripping phase were integrated in one single module unit [99, 208]. Such a set-up allowed the continuous removal of the product from the organic solvent to the stripping solution, favoring the equilibrium shift to the acid production [99]. However, due to the high cost of the modules, the production was discontinued. Instead, two hollow-fiber membrane extractors in series could be used, as demonstrated by Jin and Yang [100] in the extractive fermentation of propionic acid. The first polypropylene hollow-fiber module was used for propionic acid extraction from the fermentation broth, whereas in a second similar module the acid was stripped from the organic phase into a basic solution, direct contact between organic and aqueous phases being prevented throughout the whole process. Increases in yield, purity, final product concentration, and productivity were observed, and related to decreased product inhibition and also to possible shifts in the metabolic pathway, leading to a decreased production of secondary compounds.

A simple, less costly, experimental set-up was effectively developed by León and co-workers to carry out the oxidation of methylpropanediol in a membrane-assisted three-phase system [99]. In this system, the product was extracted from the bioconversion medium recirculated in the lumen side, to the organic phase flowing in the shell side of the membrane module. The product was then back-extracted from the organic phase to a stripping alkaline aqueous phase in a stirred tank reactor. Slow stirring prevented phase mixing and therefore recirculation of the stripping solution into the membrane reactor.

4.3 Solvent Toxicity Prevention

The advantageous use of membranes in the removal of toxic compounds produced during fermentation has been demonstrated in several fermentation systems. Increased (2.5 fold) volumetric productivities and product concentration yields (fourfold) were reported by Christen et al. [209] in ethanolic fermentations assisted by supported liquid membranes. Increased final product concentrations and yield were also reported by Xavier et al. [134] when comparing membrane-assisted extractive fermentation of lactic acid with conventional fermentations.

The effect of detoxification of the medium by removal of toxic compounds with UF membranes was demonstrated by Boyaval et al. [36] in the fermentation of propionic acid. UF runs led to an eightfold increase in volumetric productivity relative to fed batch experiments. The effectiveness of membrane bioreactors in the lowering of toxicity of the compounds involved in the bioconversion system was demonstrated by Edwards and co-workers [159]. An eightfold increase in the removal of phenolic compounds from effluents was observed when polyphenoloxidase was immobilized in a capillary poly(ether)sulfone membrane as compared to the use of the free enzyme. Butanol recovery from the fermentation medium with organic solvent extraction or membrane solvent extraction led to similar results, both processes leading to decreased product inhibition. Due to the low toxicity of the extractive solvent used (isopropyl myristate) on *Clostridium beyerinckii* cells, no protective effect of the membrane was observed. However, precipitates observed in two-liquid phase extraction were not observed when a membrane was used. On the other hand, the use of a hollow-fiber module as a separator in an organic-aqueous extractive fermentation of lactic acid allowed the use of a mixture of Alamine 336 and oleyl alcohol (40:60). This proved toxic to the cells in a direct contact two-phase system, while leading to the highest extraction efficiency [207].

5 Trends in the Development of Membrane-Assisted Extractive Bioconversions

The use of membranes provides a way to overcome the limitation of direct contact organic aqueous two-liquid phase systems, namely emulsion formation and solvent toxicity, and enable continuous operation. A better understanding of the phenomena relating phase breakthrough and the physical properties of the wetting and non-wetting phase, as well as on the fouling and concentration polarization mechanisms, allows for the development of a rational selection of the extraction system set-up. Knowledge of the kinetics and mass transfer characteristics of the bioconversion system is essential for modeling the reaction system and thus provides a tool to predict its performance. The use of membrane reactors for the field of wastewater treatment has also been increasing, with research efforts focusing on the development of effective modeling of microbial behavior and substrate degradation.

The use of thin, dense polymeric membranes is finding widespread use for their ability to prevent phase breakthrough. Ceramic membranes are also being increasingly used in organic-aqueous systems, due to their higher mechanical stability and enlarged possibility of regeneration. Development of novel polymeric materials showing enantioselectivity will allow further developments in the selective production of optically active compounds. Nanofiltration membranes, allowing the retention of co-factors, further enlarge the range of bioconversions that can be performed in a continuous manner. Also, the integration of charged membranes in perextraction systems might increase process selectivity.

The use of membrane modules for the immobilization of mammalian cells has been revived, also with focus on the development of effective predictive models for growth and production behavior.

6 References

1. Demain AL (2000) Tibtech 18:26
2. Austin GT (1986) Shreve's chemical process industries, 5th edn. McGraw-Hill, Singapore
3. Shuler ML, Kargi F (1992) Bioprocess engineering. Prentice Hall, New Jersey
4. Hogg, JA (1992) Steroids 57:593
5. Minjnbeek G, Leach CK (1997) An introduction to biotechnological innovations in the chemical industry. In: Biotechnological innovations in chemical synthesis. Butterworth-Heinemann, Oxford
6. Tramper J (1996) Biotechnol Bioeng 52:290
7. Semenov AN, Khmelnitski YL, Berezin IV, Martinek K (1987) Biocatalysis 1:3

8. Gibbs DF (1983) Tibtech 1:12
9. Minier M, Grateloup R, Blanc-Ferras E, Goma G (1990) Biotechnol Bioeng 35:861
10. Lilly MD (1994) Trans I Chem Eng 72:27
11. Walsh PK, Liu CP, Findley ME, Liapis AI, Siehr DJ (1984) Biotechnol Bioeng Symp 13:629
12 Cysewski GR, Wilke CR (1977) Biotechnol Bioeng 19:1125
13. Minier M, Goma G (1982) Biotechnol Bioeng 24:1565
14. Lencki RW, Robinson CW, Moo-Young M (1983) Biotechnol Bioeng Symp 13:617
15. Fabre CE, Condore JS, Marty A (1999) Biotechnol Bioeng 64:392
16. Freeman A, Woodley JM, Lilly MD (1993) Bio/Technology 11:1007
17. Ramalingham A, Finn RK (1977) Biotechnol Bioeng 19:583
18. Ghose TK, Roychoudhury PK, Ghose P (1984) Biotechnol Bioeng 26:377
19. Silva FLH, Rodrigues MI, Maugeri F (1999) J Chem Technol Biotechnol 74:176
20. O'Brien DJ, Craig JC (1996) Appl Microbiol Biotechnol 44:699
21. Geng Q, Park CH (1994) Biotechnol Bioeng 43:978
22. Kargupta K, Datta S, Sanyal SK (1998) Biochem Eng J 1:31
23. Dale MC, Okos MR, Wankat PC (1984) Biotechnol Bioeng 26:377
24. Qureshi N, Maddox IS (1991) Bioprocess Eng 6:63
25. Kim C-H, Kim S-W, Hong S-I (1999) Process Biochem 35:485
26. Held M, Schmid A, Kohler H-PE, Suske W, Witholt B, Wubbolts MG (1999) Biotechnol Bioeng 62:641
27. Chauhan RP, Woodley JM, Powell LW (1997) Biotechnol Bioeng 56:345
28. Lynch RM, Woodley JM, Lilly MD (1997) J Biotechnol 58:167
29. Vincenzi JT, Zmijewski MJ, Reinhard MR, Landen BE, Muth WL, Marker PG (1997) Enzyme Microb Technol 20:494
30. Zihao W, Kefeng Z (1995) Biotechnol Bioeng 47:1
31. Lee C-Y, Chen CD, Liu W-H (1993) Appl Microbiol Biotechnol 38:447
32. Corry JP, Reed WL, Curtis WR (1993) Biotechnol Bioeng 42:503
33. Shity H, Bar R (1992) Biotechnol Bioeng 39:462
34. Ohaishi R, Yamamoto T, Suzuki T (1999) J Biosci Bioeng 87:647
35. Park BG, Lee WG, Chang YK, Chang HN (1999) Bioprocess Eng 21:97
36. Boyaval P, Corre C, Madec ML (1994) Enzyme Microb Technol 16:883
37. Bryjak J, Bryjak M, Noworyta A (1996) Enzyme Microb Technol 19:196
38. Foda MI, Lopez-Leiva M (1999) Process Biochem 35:581
39. Masuda M, Kawase Y, Kawase M (2000) J Biosci Bioeng 89:119
40. Kulkarni N, Vaidya A, Rao M (1999) Biochem Biophys Res Commun 255:274
41. Planas J, Radstom P, Tjerneld F, Hahn-Haegerdal B (1996) Appl Microbiol Biotechnol 45:737
42. Umakoshi H, Yano K, Kuboi R, Komasawa I (1996) Biotechnol Prog 12:51
43. Chen JP, Lee MS (1995) Enzyme Microb Technol 17:1021
44. Lee IY, Volm TG; Rosazza JPN (1998) Enzyme Microb Technol 23:261
45. Wieczorek S, Brauer H (1998) Bioprocess Eng 18:75
46. Molinari F, Gandolfi R, Aragozzini F, León R, Prazeres DMF (1999) Enzyme Microb Technol 25:729
47. Mathys RG, Schmid A, Witholt B (1999) Biotechnol Bioeng 64:459
48. Zigova J, Sturdik E, Vandak D, Schlosser S (1999) Process Biochem 34:835
49. Ishizaki A, Michiwaki S, Crabbe E, Kobayashi G, Sonomoto K, Yoshino S (1999) J Ferment Bioeng 87:352
50. Gu Z, Glatz BA, Glatz CE (1998) Biotechnol Bioeng 57:454
51. Wubbolts MG, Hoven J, Melgert B, Witholt B (1994) Enzyme Microb Technol 16:887
52. Daugulis AJ, Axford DB, Ciszek B, Malinowski JJ (1994) Biotechnol Lett 16:637
53. Buitelaar RM, Langenhoff AAM, Heidstra R, Tramper J (1991) Enzyme Microb Technol 13
54. Honda H, Toyama Y, Takahashi H, Nakazeko T, Kobayashi T (1995) J Ferment Bioeng 79:589
55. Paradkar VM, Dordick JS (1994) Biotechnol Bioeng 43:529

56. Weilnhammer C, Blass E (1994) Chem Eng Technol 17:365
57. Bernard P, Barth D (1996) Biocatalysis Bioconversions 12:299
58. Maiorella BL, Blanch, HW, Wilke, CR (1984) Biotechnol Bioeng 26:1003
59. Daugulis AJ, Swayne DE, Kollerup F, Groom CA (1987) Biotechnol Lett 9:425
60. Bruce LJ, Daugulis AJ (1991) Biotechnol Prog 7:116
61. Patel RN (ed) (2000) Stereoselective biocatalysis. Marcel Dekker, New York
62. Nikolova P, Ward OP (1993) J Ind Microbiol 12:76
63. Sikkema J, de Bont JAM, Poolman B (1995) Microbiol Rev 59:201
64. Salter GJ, Kell DB (1995) Crit Rev Biotechnol 15:139
65. Halling PJ (1994) Enzyme Microbiol Technol 16:178
66. Klibanov AM (1997) Tibtech 15:97
67. León R, Fernandes P, Pinheiro HM, Cabral JMS (1998) Enzyme Microb Technol 23:483
68. Isken S, de Bont JAM (1998) Extremophyles 2:229
69. de Bont JAM (1998) Tibtech 16:493
70. Ramos JL, Duque E, Rodriguez-Herva JL, Godoy P, Haïdour A, Reyes F, Fernandez-Barrero A (1997) J Biol Chem 272:3887
71. Ogino H, Miyamoto K, Yasuda M, Ishimi K, Ishikawa H (1999) Biochem Eng J 4:1
72. Ahmad S, Johri N (1992) Indian J Microbiol 32:489
73. Aono R, Kobayashi H (1997) Appl Environ Microbiol 63:3637
74. Kieboom J, Dennis JJ, de Bont JAM, Zylstra GJ (1998) J Biol Chem 273:85
75. Li X-Z, Zhang L, Poole K (1998) J Bacteriol 180:2987
76. Paje MLF, Neilan BA, Couperwhite I (1997) Microbiol 143:2975
77. Ortega-Calvo J-J, Alexander M (1994) Appl Environ Microbiol 60:2643
78. Timmis KN, Piepar DH (1999) Tibtech 17:201
79. Vermuë M, Sikkema J, Verheul A, Bakker R, Tramper J (1993) Biotechnol Bioeng 42:747
80. Rajagopal AN (1996) Biotechnol Bioeng 19:606
81. Marták J, Rosenberg M, Schlosser S, Kristofiková L (1995) Biotechnol Tech 9:247
82. Rekker RF, Mannhold R (1992) In: Rekker RF, Mannhold R (eds) Calculation of drug lipophilicity. The hydrophobic fragmental constant approach. VCH, Mannheim
83. Aono R, Doukyu N (1996) Biosci Biotech Biochem 60:1146
84. Oliveira AC, Cabral JMS (1991) J Chem Technol Biotechnol 52:219
85. Minier M, Goma G (1982) Biotechnol Bioeng 24:1565
86. Minier M, Goma G (1981) Biotechnol Lett 3:405
87. Davison BH, Thompson JE (1993) Appl Biochem Biotechnol 39/40:415
88. Qureshi N, Maddox IS (1995) J Ferment Bioeng 80:185
89. Eckert G, Shugerl K (1987) Appl Microbiol Biotechnol 27:221
90. Roffler SR, Blanch HW, Wilke CR (1987) Bioprocess Eng 5:203
91. Grobben NG, Eggink CF, Cuperus FP, Huizing HJ (1993) Appl Microbiol Biotechnol 39:494
92. Ishizaki A, Michiwaki S, Crabbe E, Kobayashi G, Sonomoto K, Yoshino S (1999) J Bioscience Bioeng 87:352
93. Lee TM, Ishizaki A, Yoshino S, Furukawa K (1995) Biotechnol Lett 17:649
94. Kertes AS, King CJ (1986) Biotechnol Bioeng 28:269
95. Yang ST, White SA, Hsu HT (1991) Ind Eng Chem Res 30:1335
96. Choudhury B, Basha A, Swaminathan T (1998) J Chem Technol Biotechnol 72:111
97. Mathys RG, Kut OM, Witholt B (1998) J Chem Technol Biotechnol 71:315
98. Roffler SR, Randolph TW, Miller DA, Blanch HW, Prausnitz JM (1991) Extractive bioconversions with nonaqueous solvents. In: Mattiasson B, Holst O (eds) Extractive bioconversions. Marcel Dekker, New York, p 133
99. León R, Prazeres DMF, Fernandes P, Molinari F, Cabral JMS (2001) Biotechnol Prog 17:468
100. Jin Z, Yang S-T (1998) Biotechnol Prog 14:457
101. Yi Q, Sarney DB, Khan JA, Vulfson EN (1998) Biotechnol Bioeng 60:385
102. Xin J-Y, Li S-B, Xu Y, Wang L-L (2000) Biotechnol Bioeng 68:73
103. Eykamp W (1997) Membrane separations in downstream processing. In: Handbook of downstream processing. Chapman & Hall, London, p 90
104. Mateus M, Cabral JMS (1991) Bioseparation 2:279

105. Lye GJ, Woodley (1999) Tibtech 17:395
106. Cheryan M, Mehaia MA (1986) Membrane bioreactors, In: MacGregor WC (ed) Membrane separations in biotechnology. Marcel Dekker, New York, p 255
107. Lee SG, Kim HS (1993) Biotechnol Bioeng 42:737
108. Sims KA, Cheryan M (1992) Biotechnol Bioeng 39:960
109. Mountzouris KC, Gilmour SG, Grandison AS, Rastall RA (1999) Enzyme Microb Technol 24:75
110. Prazeres DMF, Cabral JMS (1994) Enzyme Microb Technol 16:738
111. Balcão VM, Malcata FX (1998) Biotechnol Bioeng 60:114
112. Rice KE, Watkins J, Hill CH Jr (1999) Biotechnol Bioeng 63:33
113. Bouwer ST, Cuperus FP, Derksen (1997) Enzyme Microb Technol 21:291
114. Furui M, Furutani T, Shibatani T, Nakamoto Y, Mori T (1996) J Ferment Bioeng 81:21
115. Serralheiro MLM, Prazeres DMF, Cabral JMS (1999) Enzyme Microb Technol 24:507
116. Chiang C-L, Tsai S-W (1992) J Chem Technol Biotechnol 54:27
117. Prazeres DMF, Garcia FAP, Cabral JMS (1993) Biotechnol Bioeng 41:761
118. Vaidya AM, Halling PJ, Bell G (1993) Biotechnol Tech 7:441
119. Rosell CM, Vaidya AM, Halling PJ (1996) Biotechnol Bioeng 49:284
120. Bossi A, Guerrera S, Righetti PG (1999) Biotechnol Bioeng 64:383
121. Isono Y, Nakajima M (2000) Biochemical Eng J 4:143
122. Pronk W, van der Burgt M, Boswinkel G, van't Riet K (1991) J Amer Oil Chem Soc 68:852
123. Reij MW, Keurentjes JTF, Hartmans S (1998) J Biotechnol 59:155
124. Doig SD, Boam AT, Leak DI, Livingston AG, Stuckey DC (1998) Biotechnol Bioeng 58:587
125. Wu D-R, Cramer SM, Belfort G (1993) Biotechnol Bioeng 41:979
126. Schroën CGP, Woodley JM (1997) Biotechnol Prog 13:276
127. Vaidya AM, Bell G, Halling PJ (1992) J Memb Sci 71:139
128. Belhocine D, Mokrane H, Grib, Lounici H, Pauss A, Mameri N (2000) Chem Eng J 76:189
129. Paolucci-Jeanjean D, Belleville MP, Rio GM, Zakhia N (2000) Biochem Eng J 6:233
130. Cass BJ, Schade F, Robinson CW, Thompson JE, Legge RL (2000) Biotechnol Bioeng 67:372
131. Nishizawa K, Nakajima M, Nabetani H (2000) Biotechnol Bioeng 68:92
132. Nakajima M, Nishizawa K, Nabetani H (1993) Bioprocess Eng 9:31
133. Bosetti A, Bianchi D, Andriollo N, Cidaria D, Cesti P, Sello G, Di Gennaro P (1996) J Chem Technol Biotechnol 66:375
134. Xavier AMRB, Gonçalves LMD, Moreira LMD, Carrondo MJT (1995) Biotechnol Bioeng 45:320
135. Liese A, Zelinski T, Kula M-R, Kierkels H, Karutz M, Kragl U, Wandrey (1998) J Mol Catal B: Enzymatic 4:91
136. Fujita M, Era A, Ike M, Soda S, Miyata N, Hirao T (2000) J Biosci Bioeng 90:387
137. Lloyd JR, Harding CL, Macaskie LE (1997) Biotechnol Bioeng 55:505
138. Nidetzky B, Haltrich D, Schmidt K, Schmidt H, Weber A, Kulbe KD (1996) Biocatalysis Biotransform 14:47
139. Lin S-S, Harada T, Hata C, Miyawaki O, Nakamura K (1997) J Ferment Bioeng 83:54
140. Kim DM, Choi CY (1996) Biotechnol Prog 12:645
141. Bódalo A, Gómez JL,Gómez E, Bastida J, Máximo MF, Montiel (2001) Enzyme Microb Technol 28:355
142. Widjaja A, Ogino H, Yasuda M, Ishimi K, Ishikawa H (1999) J Biosci Bioeng 88:640
143. Chuichulcherm S, Nagpal S, Peeva L, Livingston A (2001) J Chem Technol Biotechnol 76:61
144. Ferreira Jorge RM, Livingston AG (2000) Biotechnol Bioeng 70:313
145. Yamamoto Y, Sugimoto S, Shen X-C, Nagamune T, Yao S-I, Suzuki E (1999) Biochem Eng J 3:151
146. Xiao Z, Dahlan MH, Xing X-H, Yoshikawa Y, Matsumoto K (2000) Biochemical Eng J. 5:83
147. Edwards W, Bownes R, Leukes WD, Jacobs EP, Sanderson R, Rose PD, Burton SG (1999) Enzyme Microb Technol 24:209
148. Doig SD, Boam AT, Livingston AG, Stuckey DC (1999) Biotechnol Bioeng 63:601

149. Isono Y, Fukushima K, Kawakatsu T, Nakajima M (1997) Biotechnol Bioeng 56:162
150. Westgate S, Vaidya AM, Bell G, Halling PJ (1998) Enzyme Microb Technol 22:575
151. Choi WJ, Choi CY, de Bont JAM, Weijers CAGM (1999) Appl Microbiol Biotechnol 53:7
152. Giorno L, Molinari R, Drioli E, Bianchi P, Cesti P (1995) J Chem Tech Biotechnol 64:345
153. Merçon F, SantÁnna GL Jr, Nobrega (2000) J Am Oil Chem Soc 77:1043
154. Xu X, Balchen S, Jonsson G, Adler-Nissen J (2000) J Am Oil Chem Soc 77:1035
155. Molinari F, Aragozzini F, Cabral JMS, Prazeres DMF (1997) Enzyme Microb Technol 20:604
156. Isono Y, FukushimaK, Araya G, Nabetani H, Nakajima M (1997) J Chem Technol Biotechnol 70:171
157. Vaidya AM, Bell G, Halling PJ (1994) J Memb Science 13:26
158. Reiken SR, Knob RJ, Briedis DM (1990) Enzyme Microb Technol 12:736
159. Edwards W, Bownes R, Leukes WD, Jacobs, Sanderson R, Rose PD, Burton SG (1998) Enzyme Microb Technol 24:209
160. Mulder M (1992) Basic principles of membrne technology. Kluwer Academic, Dordrecht
161. Jirage KB, Martin CR (1999) Tibtech 17:197
162. Rios GM, Joulie R, Sarrade SJ, Carles M (1996) AIChE J 44:2521
163. Seelbach K, Kragl U (1997) Enzyme Microb Technol 20:389
164. Tsuru T, Shutou T, Nakao S-L, Kimura S (1994) Sep Biotechnol Tech 29:971
165. Timmer JMK, Kromkamp J, Robbertsen T (1994) J Memb Sci 92:185
166. Nidetzky B, Neuhauser W, Haltrich D, Kulbe KD (1996) Biotechnol Bioeng 52:387
167. Seader JD, Henley EJ (1998) Separation process principles. Wiley & Sons, New York
168. Cussler EL (1997) Diffusion – mass transfer in fluid systems, 2nd edn. Cambridge University Press, Cambridge
169. Ogata N (1997) Macromol Symp 118:693
170. Liu Y, Zhao M, Bergbreiter DE, Crooks RM (1997) J Am Chem Soc 119:8720
171. Mikulasek P, Wakeman RJ, Merchant JQ (1999) Chem Biochem Eng Q 13:73
172. Wisniewski C, Cruz AL, Grasmick A (1999) Biochemical Eng J 3:61
173. Leonard D, Mercier-Bonin M, Lindley ND, Lafforgue C (1998) Biotechnol Prog 14:680
174. Tsuru T, Takezoe H, Asaeda M (1998) AIChE J 44:765
175. Heath CA, Belfort G (1992) Adv Biochem Eng/Biotechnol 47:46
176. Ho WSW, Sirkar KK (1992) Membrane handbook. Van Nostrand Reinhold, New York
177. Zhang W, Park BG, Chang YK, Chang HN, Yu XJ, Yuan Q (1998) Bioprocess Eng 18:317
178. Hakoda M, Enomoto A, Hoshino T, Shiragami N (1996) J Ferment Bioeng 82:361
179. Cook MA (1996) Membrane processing in biotechnology: a case history. In: Verrall MS (ed) Downstream processing of natural: a practical handbook. Wiley & Sons, New York
180. Gaeta SN (1995) The industrial development of polymeric membranes and membrane modules for reverse osmosis and ultrafiltration. In: Caetano A, Pinho MN, Drioli E, Muntau (eds) Membrane technology: applications to industrial wastewater treatment. Kluwer Academic, Dordrecht, p 25
181. Harscoat C, Jaffrin MY, Paullier P, Courtois B, Courtois J (1999) J Chem Technol Biotechnol 74:571
182. Hoq MM, Yamane T, Shimizu S, Funada T, Ishida S (1984) J Am Oil Chem Soc 61:776
183. Pronk W, Kerkhof PJAM, van Helden C, van't Riet K (1988) Biotechnol Bioeng 32:518
184. Vaidya AM, Halling PJ, Bell G (1994) Biotechnol Bioeng 44:765
185. Solichien MS, O'Brien D, Hammond EG, Glatz CE (1995) Enzyme Microb Technol 17:23
186. Anonymous (1995) Mycrodyn-modules with PP-membranes, technical information, mycrodyn modulblau. GmbH & Co., Wuppertal
187. Schroën CGPH, Cohen Stuart MA, van der Padt MA, Van't Riet, K (1994) Bioseparation 4:151
188. Conrad PB, Lee SS (1998) Biotechnol Bioeng 57:631
189. Prasad R, Sirkar KK (1992) Membrane-based solvent extraction. In: Ho WSW, Sirkar KK (eds) Membrane handbook. Van Nostrand Reinhold, New York, p 727
190. Matsumura M (1991) Perstraction. In: Mattiasson B, Holst O (eds) Extractive bioconversions. Marcel Dekker, New York, p 91

191. Gabelman A, Hwang S-T (1999) J Memb Sci 159:61
192. Pressman JG, Georgiou G, Speitel GE Jr (1999) Biotechnol Bioeng 62:681
193. Dahuron L, Cussler EL (1988) AIChE J 34; 130
194. Yang MC, Cussler EL (1986) AIChE J 32:1910
195. Pronk W, Boswinkel G, van't Riet K (1992) Enzyme Microb Technol 14:214
196. Malcata FX, Hill GC, Amundson CH (1991) Biotechnol Bioeng 38:853
197. Wu D-R, Belfort G, Cramer SM (1990) Ind Eng Chem Res 29:1612
198. Sehanoputri PS, Hill CG Jr (2000) Biotechnol Bioeng 69:450
199. Giorno L, Drioli E (1997) J Chem Technol Biotechnol 69:11
200. Rindfleisch D, Syska B, Lazarova Z, Schuegerl K (1997) Process Biochem 32:605
201. Isono Y, Nabetani H, Nakajima M (1995) Process Biochem 30:773
202. Carvalho CML, Aires-Barros MR, Cabral JMS (1999) Biotechnol Bioeng 66:17
203. Wehtje E, Swensson I, Adlercreutz P, Mattiasson B (1993) Biotechnol Tech 7:873
204. Liesse A, Filho MV (1999) Curr Opin Biotechnol 10:595
205. Hashimoto S, Ozaki (1999) Curr Opin Biotechnol 10:604
206. Pressman JG, Georgiou G, Speitel GE Jr (2000) Biotechnol Bioeng 68:548
207. Ye K, Jin S, Shimizu K (1996) J Ferment Bioeng 81:240
208. Majumdar S, Sirkar KK (1992) Hollow-fiber contained liquid membrane. In: Ho WSW, Sirkar KK (eds) Membrane handbook. Van Nostrand Reinhold, New York, p 764
209. Christen P, Minier M, Renon H (1990) Biotechnol Bioeng 36:116

Received: February 2002

In Situ Product Removal (ISPR) in Whole Cell Biotechnology During the Last Twenty Years

Daniel Stark · Urs von Stockar

Laboratory of Chemical and Biochemical Engineering, Swiss Federal Institute of Technology (EPFL), 1015 Lausanne, Switzerland. *E-mail: urs.vonstockar@epfl.ch*

This review sums up the activity in the field of in situ product removal in whole cell bioprocesses over the last 20 years. It gives a complete summary of ISPR operations with microbial cells and cites a series of interesting ISPR applications in plant and animal cell technology. All the ISPR projects with microbial cells are categorized according to their products, their ISPR techniques, and their applied configurations of the ISPR set-up. Research on ISPR application has primarily increased in the field of microbial production of aromas and organic acids such lactic acid over the last ten years. Apart from the field of de novo formation of bioproducts, ISPR is increasingly applied to microbial bioconversion processes. However, despite of the large number of microbial whole cell ISPR projects (approximately 250), very few processes have been transferred to an industrial scale. The proposed processes have mostly been too complex and consequently not cost effective. Therefore, this review emphasizes that the planning of a successful whole cell ISPR process should not only consider the choice of ISPR technique according to the physicochemical properties of the product, but also the potential configuration of the whole process set-up. Furthermore, additional process aspects, biological and legal constraint need to be considered from the very beginning for the design of an ISPR project. Finally, future trends of new, modified or improved ISPR techniques are given.

Keywords. In situ product removal (ISPR), Integrated bioprocessing, Whole cell bioprocesses

Advances in Biochemical Engineering/
Biotechnology, Vol. 80
Series Editor: T. Scheper

1 Introduction

Compared to chemical processes, a biotechnological process using whole cell biocatalysts is characterized by low productivity because of inhibition at product concentrations. Furthermore, the product stream is dilute, which leads to high costs in the subsequent isolation and purification of the product. The performance of biotechnological processes can be enhanced by either strain improvements (screening of mutants, recombinant DNA technology) or by process engineering solutions. Although considerable progress has been achieved by the former measure, improvements of the production system still need to be applied [1].

Besides improving oxygen input and heat transfer in the reaction system or optimizing single downstream processing steps, the most common approach to raise the productivity of a fermentation process is to increase the cell concentration in the fermenter. A high-density culture can be achieved either by immobilization of the biocatalyst in the reactor or by retaining the cells in the fermenter via cell recycling using membrane filtration [2]. Thereby, substrate concentration in the reactor is controlled by a continuous feed. It needs to be stressed that this method increases the volumetric productivity, but it yields a dilute product stream. In addition, it does not remove product inhibition, as the yield of product per consumed substrate does not change by draining.

Another approach to increase the productivity of a biotechnological process is to remove the inhibitory product from the vicinity of the biocatalyst as soon as it is formed. This in situ product removal (ISPR) can increase the productivity or yield of a given biological process by any of the following means [3]: a) overcoming inhibitory or toxic effects of product to allow continuous formation at maximal production level, b) minimizing product losses owing to degradation or uncontrolled release (e.g., by evaporation), and c) reducing the total number of downstream processing steps. ISPR is restricted to extracellular products, since it is very difficult to release intracellular products without affecting cell viability [4]. Intracellular products from microbial cells are separated after the cell mass is destroyed. Furthermore, ISPR is also applied to remove byproducts such as ethanol or lactic acid that lower the performance of a fermentation process. ISPR, often synonymously called "extractive" fermentation or bioconversion, is part of the general idea of integrated bioprocessing, which represents the general coordination of upstream, reaction and downstream technologies.

Preliminary research on the application of ISPR techniques in biotechnology was done in the 1960s and 1970s for the on-line removal of a toxin by an aqueous two-phase system [5], salicylic acid by ion-exchange resins [6], 7,8-epoxy-1-octene by extraction into a water-immiscible solvent [7], and ethanol by vacuum fermentation [8]. Dialysis was applied to remove lactic acid [9] and cycloheximide [10]. Large scale fermentative production of lactic and citric acid was done by the addition of lime to precipitate the calcium carboxylates [11]. The downstream processing required the acidification of the carboxylates by sulfuric acid, which resulted in the stoichiometric coproduction of gypsum. Thorough research on the different ISPR techniques began in the early 1980s to raise the productivity of ethanol fermentation with respect to ethanol production in the petrochemical industry. Based on the volatility and the hydrophobicity of this solvent many different ISPR techniques were investigated. However, none of the proposed set-ups was realized on an industrial scale apart from the BIOSTIL process [12]. Yeast is concentrated in the fermenter by cell recycling through a centrifuge [13] in this continuous process. Furthermore, the inhibitory ethanol is separated in a distillation column in the external loop. The BIOSTIL process is essentially a process that effectively recycles water and thus reduces equipment size. Each unit operation includes robust and well-tested equipment, ensuring the industry's high demands on reliability of operation. In addition to the application of ISPR to the ethanol fermentation, much effort has also been reported in the production of the solvents butanol and acetone and several organic acids such as lactic and butyric acid. A considerable number of ISPR articles are also available on steroid conversions, aroma compounds, secondary metabolites, and various fine chemicals.

A significant number of general reviews on ISPR techniques in whole cell bioprocessing have been published [1, 3, 14–21]. Furthermore, more specialized reviews exist that cover either a certain product category such as ethanol [4] or butanol [22, 23] or the use of certain specialized ISPR techniques such as aqueous two-phase system [24, 25], organic-aqueous two-phase systems [26–30], or solid adsorbents [31].

Most of these reviews propose, on the basis of the physicochemical properties of the target product, a systematic approach for the selection of appropriate ISPR techniques. As a result, all the authors have found several possible ISPR techniques that are able to remove the product selectively from the reaction mixture and consequently are able to increase the productivity of the process. In addition, most of the theoretical reflections of the researchers have even been proven on a laboratory scale. Hence, it is justified to elucidate the reasons why almost none of these processes were transferred to an industrial scale. ISPR projects of the last twenty years that have been using microbial cells are summarized in this article. They are categorized according to their product category and reactor set-up. Additional constraints that influence the success of an ISPR implementation are also discussed. This summary gives a basis for the discussion of some of the past and present trends in ISPR applications. Furthermore, future activities and new ISPR techniques are briefly discussed. Finally, several promising ISPR applications in the growing field of plant and animal cell technology are shown.

2 Matching the Appropriate ISPR Techniques to Different Product Categories

A lot of attention has been paid to the choice of possible ISPR techniques according to the physicochemical properties of the target product. The aforementioned reviews propose systematic approaches to select successful methods to remove the target product from the vicinity of the cell. ISPR is therefore designed and affected via exploitation of the difference in molecular properties of the product relative to the background medium. Freeman and coworker proposed five principal product properties to help choose the most suitable ISPR techniques [3, 21]. Volatility (boiling point $<80\,°C$), hydrophobicity ($\log P_{oct} > 0.8$), size (molecular weight <1000 Da), charge (positive, negative, neutral), and specific binding properties of a compound can be used to group and assign the products to their appropriate ISPR methods (Table 1).

A product may be removed from its producing cell by five main possible techniques. Evaporation occurs via stripping, (vacuum) distillation or by membrane-supported techniques such as pervaporation and transmembrane distillation. Extraction into another phase includes the use of water-immiscible organic solvents, supercritical fluids, or an aqueous two-phase system. In addition, the

Table 1. Appropriate ISPR techniques for different product categories. (+) technique is applicable for selective removal. (++) technique is very useful for selective removal of the product and has shown its effectiveness in several cases

ISPR Technique	Product categories Low molecular wt					High molecular wt			Important examples
	Hydrophobic – volatile	Hydrophobic – nonvolatile	Hydrophobic – neutral – volatile	Hydrophobic – neutral – nonvolatile	Hydrophobic – charged	Hydrophobic	Neutral	Charged	
Evaporation									
stripping, (vacuum-) distillation, per vaporation, transmembrane distillation	++		++						ethanol
Extraction									
organic solvent (inc. perstraction)	++	++	+	+		+			ethanol, flavors
supercritical fluid	+	++	+	+	+				flavors
reactive (inc. perstraction)	+	+	+	+	++	+	+	+	organic acids, ethanol
aqueous two-phase			+	+	+	+	++	++	enzymes
Permeation									
dialysis			+	+	+				lactic acid,ethanol, ammonium
electrodialysis					++				lactic acid
reverse osmosis, nanofiltration			+	+	+				ethanol
Immobilization									
hydrophobic adsorption	++	++	+	+		+	+	+	butanol, flavors
ion-exchange					++			+	organic acids
affinity adsorption	+	+	+	+	+	++	++	++	proteins
Precipitation					+			+	lactic acid, citric

second phase can include a reagent that complexes for instance organic acids or esterifies alcohols. The techniques including an organic phase can be supported by a membrane (perstraction). Size selective permeation techniques such as dialysis, electrodialysis, reverse osmosis, or nanofiltration take advantage of membranes. Immobilization procedures include the adsorption on hydrophobic carriers, affinity adsorption techniques on the basis of molecular recognition, and ion-exchange resins. Finally, certain 'lucky' cases exist in which the charged product can be precipitated by a counter-ion during the fermentation. There are often a variety of techniques available to remove a specific product.

3
Categorization of Microbial Cell ISPR During the Last Twenty Years

ISPR in biotechnology is applied to whole cell and enzymatic biocatalysis. However, the application of ISPR techniques in enzymatic biocatalysis is not discussed in this publication. This review focuses on ISPR applied to whole cells and gives a complete summary of all work published either in literature or patents within last twenty years. It also includes the key articles that were published before 1980. The following points explain the criteria and restrictions that have been used for the selection of the listed ISPR projects.

- The review completely covers only ISPR applications with microbial cells. Some interesting applications using animal and plant cells are given in the outlook section.
- The summary is based on the number of projects of the different research groups and not their number of publications. Only the original publication is taken into account if several articles have been published showing the same results. ISPR projects on the same bioconversion using different configurations of the separation technique are counted independently.
- The review includes only articles that show experimental results of a bioprocess with a simultaneous removal of the product. Publications that independently cover a certain separation technology and only mention its possible application for an ISPR process are not taken into account.
- Only articles that raise the productivity of a bioprocess by removing the product selectively are included. Therefore, processes that improve their productivity by the implementation of a simple cell recycle or another immobilization system are not taken into account. Nor is dialysis, which is mainly used as a biomass retention system or sometimes as a means of removing unwanted by-products. Information about dialysis cultures is available in the review by Pörtner and Märkl [32].

All the projects are categorized in Table 2 according to their product (category) species and sorted by the year of publication. The applied ISPR techniques of the different projects are grouped according the categories introduced in the table. All utilized strains are listed and it is indicated if the product was formed through a bioconversion or from a de novo synthesis. The numbers refer to the different ISPR configurations depicted in Fig. 1 and are discussed below.

Table 2. ISPR projects in microbial whole cell biotechnology

Product category	Product	Strain		Extraction						Immobilis.			Evaporation				Perme-ation			Year	Reference
			Denovo (dn), bioconv. (b)	Organic solvent	Reactive	Aqeuous two-phase	Perstraction (org. phase)	Reactive perstraction	Supercritical fluids	Ion-exchange	Hydrophobic adsorption	Affinity adsorption	Stripping	Distillation (inc. vacuum)	Transmembrane distillation	Pervaporation	Electrodialysis	Nanofiltrat., rev. osmosis	Precipitation		
solvents	ethanol	*Saccharomyces cerevisiae*	dn											1						1977	[8]
	ethanol	*Saccharomyces cerevisiae*	dn											1						1977	[50]
	ethanol	*Saccharomyces cerevisiae*	dn											1						1980	[51]
	ethanol	*Zymomonas mobilis*	dn			1														1980	[52]
	ethanol	*Zymomonas mobilis*	dn											3						1981	[53]
	ethanol	*Saccharomyces cerevisiae*	dn			1														1981	[54]
	ethanol	*Saccharomyces cerevisiae*	dn	4c							4c									1981	[55]
	ethanol	*Saccharomyces cerevisiae*	dn								3									1982	[56]
	ethanol	*Saccharomyces cerevisiae*	dn	1																1982	[57]
	ethanol	*Saccharomyces cerevisiae*	dn				4b						4b							1983	[58]
	ethanol	*Saccharomyces cerevisiae*	dn								3									1983	[59]
	ethanol	*Saccharomyces cerevisiae*	dn										1							1983	[60]
	butanol	*Clostridium acetobutylicum*	dn								3									1984	[61]
	butanol, *i*-propanol	*Clostridium beyerinckii*	dn													4b				1984	[62]
	ethanol	*Saccharomyces cerevisiae*	dn	2a																1984	[63]
	ethanol	*Saccharomyces cerevisiae*	dn			3														1984	[64]
	ethanol	*Saccharomyces cerevisiae*	dn								1									1984	[65]
	ethanol	*Candida acidothermophilum*	dn											3						1984	[66]
	acetone, butanol	*Clostridium acetobutylicum*	dn	1							3									1985	[67]
	ethanol	*Kluyveromyces marxianus*	dn										2a							1985	[68]
	ethanol	*Pachysolen tannophilus*	dn								3									1985	[69]
	ethanol	*Saccharomyces cerevisiae*	dn										1							1985	[70]
	butanol, *i*-propanol	*Clostridium beyerinckii*	dn								1									1986	[71]
	butanol	*Clostridium acetobutylicum*	dn															4c		1986	[72]
	ethanol	*Saccharomyces cerevisiae*	dn	2a																1986	[73]
	ethanol	*Saccharomyces cerevisiae*	dn	2a																1986	[74]
	ethanol	*Saccharomyces cerevisiae*	dn				4b													1986	[75]
	ethanol	*Saccharomyces cerevisiae*	dn				4a													1986	[76]
	ethanol	*Saccharomyces cerevisiae*	dn															4c		1986	[77]
	acetone, butanol	*Clostridium acetobutylicum*	dn	1																1987	[78]
	acetone, butanol	*Clostridium acetobutylicum*	dn	4c																1987	[79]
	acetone, butanol, ethanol	*Clostridium acetobutylicum*	dn								4b		4b							1987	[80]
	butanol	*Clostridium acetobutylicum*	dn													2b				1987	[81]
	ethanol	*Candida pseudotropicalis*	dn										1							1987	[82]
	butanol	*Clostridium acetobutylicum*	dn	1																1987	[83]
	ethanol	*Saccharomyces cerevisiae*	dn	2a																1987	[84]
	ethanol	*Saccharomyces cerevisiae*	dn													4c				1987	[85]
	ethanol	*Saccharomyces cerevisiae*	dn	3																1987	[86]
	ethanol	*Saccharomyces cerevisiae*	dn	1,2 b																1988	[87]
	ethanol	*Saccharomyces cerevisiae*	dn	3																1988	[88]
	butanol	*Clostridium acetobutylicum*	dn				4a													1989	[89]
	2,3-butanediol	*Klebsiella oxytoca*	dn	3																1989	[90]
	acetone, butanol, ethanol	*Clostridium acetobutylicum*	dn				4b													1989	[91]
	butanol	*Clostridium* sp.	dn										3							1989	[92]
	ethanol	*Saccharomyces cerevisiae*	dn													4c				1989	[93]
	ethanol	*Kluyveromyces fragilis*	dn												4a					1989	[94]
	ethanol	*Saccharomyces cerevisiae*	dn													1				1989	[95]
	ethanol	*Saccharomyces cerevisiae*	dn				4b													1990	[96]
	ethanol	*Clostridium thermohydrosul.*	dn													3				1990	[97]
	ethanol	*Saccharomyces cerevisiae*	dn				4b													1990	[98]
	ethanol	*Saccharomyces bayanus*	dn													4a				1990	[99]
	ethanol	*Saccharomyces cerevisiae*	dn													4a				1990	[100]
	acetone, butanol, ethanol	*Clostridium acetobutylicum*	dn										1							1991	[101]
	acetone, butanol, ethanol	*Clostridium acetobutylicum*	dn													4b				1991	[102]
	ethanol	*Saccharomyces bayanus*	dn		2a															1991	[103]
	ethanol	*Saccharomyces cerevisiae*	dn													4b				1991	[104]
	ethanol	*Saccharomyces cerevisiae*	dn	3																1991	[105]
	ethanol	*Saccharomyces cerevisiae*	dn											4c						1991	[12]
	butanol, *i*-propanol	*Clostridium isopropylicum*	dn													4b				1992	[106]

Table 2 (continued)

Product category	Product	Strain		Extraction						Immobilis.			Evaporation				Perme-ation			Year	Reference
			Denovo (dn), bioconv. (b)	Organic solvent	Reactive	Aqeuous two-phase	Perstraction (org. phase)	Reactive perstraction	Supercritical fluids	Ion-exchange	Hydrophobic adsorption	Affinity adsorption	Stripping	Distillation (inc. vacuum)	Transmembrane distillation	Pervaporation	Electrodialysis	Nanofiltrat., rev. osmosis	Precipitation		
solvents	acetone, butanol, ethanol	*Clostridium acetobutylicum*	dn													4b				1992	[107]
	acetone, butanol	*Clostridium acetobutylicum*	dn			1														1992	[108]
	ethanol	*Saccharomyces cerevisiae*	dn													4a,c				1992	[109]
	ethanol	*Saccharomyces cerevisiae*	dn	3																1992	[110]
	ethanol	*Clostridium thermohydrosul.*	dn	1																1992	[111]
	ethanol	*Zymomonas mobilis*	dn	1																1992	[112]
	ethanol	*Saccharomyces cerevisiae*	dn				4c													1992	[113]
	ethanol	*Zymomonas mobilis*	dn								2a									1993	[114]
	acetone, butanol	*Clostridium acetobutylicum*	dn	2a																1993	[115]
	butanol	*Clostridium isopropylicum*	dn	4b																1993	[116]
	acetone, butanol	*Clostridium acetobutylicum*	dn										2a							1993	[117]
	acetone, butanol, ethanol	*Clostridium acetobutylicum*	dn				4b, c													1993	[118]
	ethanol	*Candida pseudotropicalis*	dn	1																1993	[119]
	butanol	*Clostridium acetobutylicum*	dn													2b				1994	[120]
	acetone, butanol	*Clostridium acetobutylicum*	dn				2b													1994	[121]
	ethanol	*Saccharomyces cerevisiae*	dn	1																1994	[122]
	ethanol	*Clostridium thermohydrosul.*	dn	1																1994	[123]
	acetone, butanol, ethanol	*Clostridium acetobutylicum*	dn	4b																1995	[124]
	acetone, butanol, ethanol	*Clostridium acetobutylicum*	dn	1																1995	[125]
	acetone, butanol, ethanol	*Clostridium bekjerinckii*	dn													4b				1996	[126]
	ethanol	*Saccharomyces cerevisiae*	dn													4a				1996	[127]
	ethanol	*Saccharomyces cerevisiae*	dn	4b																1996	[128]
	ethanol	*Saccharomyces cerevisiae*	dn	1																1996	[129]
	ethanol	*Saccharomyces cerevisiae*	dn		2a															1997	[130]
	ethanol	*Saccharomyces cerevisiae*	dn											4b						1997	[131]
	ethanol	*Saccharomyces cerevisiae*	dn										3							1998	[132]
	ethanol	*Saccharomyces cerevisiae*	dn		1															1998	[133]
	acetone, butanol, ethanol	*Clostridium bekjerinckii*	dn													4a				1999	[134]
	acetone, butanol, ethanol	*Clostridium saccharoperbut.*	dn	1																1999	[135]
	ethanol	*Saccharomyces cerevisiae*	dn	2a																1999	[136]
	ethanol	*Saccharomyces cerevisiae*	dn											4c						1999	[137]
	ethanol	*Pichia stipitis*	dn										1							2000	[138]
organic acids	lactic acid	*Lactobacillus delbrueckii*	dn																1	1983	[11]
	lactic acid	*Lactobacillus acidophilus*	dn														4a			1986	[139]
	acetic-, butyric acid	acidogenic bacterium	dn					4a												1987	[140]
	lactic acid	*Lactobacillus helveticus*	dn														4c			1987	[141]
	lactic acid	*Lactobacillus delbrueckii*	dn														4b			1987	[142]
	lactic acid	*Lactobacillus casei*	dn														3,4 c			1987	[143]
	lactic, acetic, citric acid	*Lactonb.,Acetobacter,A.niger*	dn		1,2 a															1987	[144]
	acetic, butyric acid	*Clostridium acetobutylicum*	dn								4c									1988	[145]
	butyric acid	*Clostridium acetobutylicum*	dn	1																1990	[146]
	lactic acid	*Lactobacillus delbrueckii*	dn														4c			1990	[147]
	lactic acid	*Lactobacillus delbrueckii*	dn		4b															1991	[148]
	lactic acid	*Lactococcus lactis*	dn														4c			1991	[149]
	lactic acid	*Lactobacillus delbrueckii*	dn		4c						4c									1991	[150]
	lactic acid	*Lactobacillus delbrueckii*	dn							3										1992	[151]
	propionic acid	*Propionibacterium acidiprop.*	dn		4b															1992	[152]
	acetic acid	*Acetogenium kivui*	dn														4c			1993	[153]
	lactic acid	*Lactobacillus casei*	dn							4c										1993	[154]
	octanoic acid	recomb. *Escherichia coli*	b	1																1993	[155]
	propionic acid	*Propionibacterium acidiprop.*	dn														4a			1993	[156]
	propionic, acetic acid	*Propionibacterium shermani*	dn														4c			1993	[157]
	lactic acid	*Lactobacillus casei*	dn							4b										1994	[158]
	lactic acid	*Lactococcus lactis*	dn														4c			1994	[159]

Table 2 (continued)

Product category	Product	Strain		Extraction						Immobilis.			Evaporation				Permeation			Year	Reference
			Denovo (dn), bioconv. (b)	Organic solvent	Reactive	Aqeuous two-phase	Perstraction (org. phase)	Reactive perstraction	Supercritical fluids	Ion-exchange	Hydrophobic adsorption	Affinity adsorption	Stripping	Distillation (inc. vacuum)	Transmembrane distillation	Pervaporation	Electrodialysis	Nanofiltrat., rev. osmosis	Precipitation		
acids	lactic acid	*Lactococcus lactis*	dn			1														1994	[160]
	acetic acid	*Propionibacterium shermani*	dn														4a			1994	[161]
	glutamic acid	*Corynebacterium glutamicum*	dn								3									1995	[162]
	gluconic acid	*Gluconobacter oxydans*	dn							3										1995	[162]
	lactic acid	*Lactobacillus casei*	dn			1														1995	[163]
	lactic acid	*Lactobacillus delbrueckii*	dn		4c															1995	[164]
	lactic acid	*Lactobacillus plantarum*	dn							4c										1995	[165]
	propionic acid	*Propionibacterium acidiprop.*	dn					4a												1995	[166]
	fumaric acid	*Rhizopus oryzae*	dn								4b									1996	[167]
	lactic acid	*Lactobacillus delbrueckii*	dn								4b									1996	[168]
	lactic acid	*Lactococcus lactis*	dn			1														1996	[169]
	lactic acid	*Lactobacillus delbrueckii*	dn		4c															1996	[38]
	lactic acid	*Lactococcus lactis*	dn			1														1996	[170]
	lactic acid	*Lactobacillus delbrueckii*	dn		4c															1996	[171]
	lactic, acetic, propionic, butyric acid	*Streptococcus, Lactobacillus, Propionibacterium*	dn				4b	4b												1996	[172]
	acetic acid	*Clostridium ljungdahlii*	dn		4c															1997	[173]
	tridecane 1,13-dicarboxylic acid	*Cryptococcus neoformans*	b								1									1997	[174]
	hexanoic acid	*Megasphaera elsdenii*	dn							1,2 a 3										1997	[175]
	lactic acid	*Lactobacillus delbrueckii*	dn					2b												1997	[176]
	citric acid	*Aspergillus niger*	dn		2b															1998	[177]
	lactic acid	*Lactobacillus delbrueckii*	dn								3									1998	[178]
	lactic acid	*Lactobacillus rhamnosus*	dn		1															1998	[179]
	lactic acid	*Lactobacillus rhamnosus*	dn		4c															1998	[180]
	lactic acid	*Lactobacillus delbrueckii*	dn								4c									1998	[181]
	octanoic acid	*Pseudomonas oleovorans*	b	1																1998	[182]
	octanoic acid	recomb. *Escherichia coli*	b	1																1998	[183]
	propionic acid	*Propionibacterium acidiprop.*	dn					4a												1998	[184]
	butyric acid	*Clostridium butyricum*	dn	1,3																1999	[34]
	lactic acid	*Lactobacillus delbrueckii*	dn							2a										1999	[185]
	lactic acid	*Lactobacillus delbrueckii*	dn					4b												1999	[186]
	propionic acid	*Propionibacterium thoenii*	dn					4c												1999	[187]
	lactic acid	*Lactobacillus casei*	dn							4c										2000	[188]
aromas	*l*-menthol	*Bacillus subtilis*	b	1																1986	[189]
	phenylethylacetate, citronellol	*Kluyveromyces marxianus*, *Ambrosiozyma monospora*	b, dn								2b									1988	[190]
	vanillyl-alcohol	*Saccharomyces cerevisiae*	b	2a																1989	[191]
	various terpens	*Hypomyces odoratus*	dn								1,2 b									1989	[192]
	benzaldehyde	*Ischnoderma benzoicum*	b										1							1990	[193]
	var. methyl-ketones	*Penicillium roquefortii*	b	1,2 a																1992	[194]
	benzylalcohol	*Saccharomyces cerevisiae*	b	1																1992	[195]
	decalactone	*Saccharomyces cerevisiae*	b								1									1992	[196]
	6-pentyl-α-pyrone	*Trichoderma viride*	dn													4a				1992	[197]
	benzaldehyde	*Pichia pastoris*	b	1,2 a																1994	[198]
	decalactone	*Tyromyces sambuceus*	b	2a																1994	[199]
	fatty acid ethyl esters	*Pseudomonas fragi*	b										1							1994	[200]
	2-phenylethanol	*Saccharomyces cerevisiae*	b	2a																1994	[201]
	benzaldehyde	*Poria cocos*	b								4c									1995	[202]
	(*S*)-citronellic acid	*Geotrichum candidum*	b				4b													1996	[203]
	benzaldehyde	*Bjerkandera adusta*	b													4a				1996	[204]
	isovaleraldehyde	*Gluconobacter oxydans*	b				4a													1997	[205]
	4-vinylguaiacol	*Bacillus pumilus*	b	1																1998	[206]
	benzaldehyde	*Pseudomonas putida*	b								3		1							1998	[207]
	citronellol	*Saccharomyces cerevisiae*	b				2b													1998	[208]
	citronellol	*Saccharomyces cerevisiae*	b	1																1998	[209]
	decalactone	*Tyromyces sambuceus*	b								2a	2a								1998	[45]
	decalactone	*Sporidiobolus salmonicolor*	b								1									1998	[210]
	geranyl acetate	*Rhizopus delemar*	b	1																1998	[211]
	vanillin	*Phanerochaete chrysosporium*	b								1									1998	[212]
	2-phenylethanol	*Kluyveromyces marxianus*	b						4c											1999	[213]
	decalactone	*Sporidiobolus salmonicolor*	b	1							1									1999	[214]
	decalactone	*Sporidiobolus salmonicolor*	b				3													1999	[215]
	wine must comp.	*Saccharomyces cerevisiae*	dn													4a				1999	[216]
	phenylacetaldehyde	*Acetobacter* sp.	b	1																1999	[217]

Table 2 (continued)

Product category	Product	Strain		Extraction						Immobilis.			Evaporation				Perme-ation			Year	Refer-ence
			Denovo (dn), bioconv. (b)	Organic solvent	Reactive	Aqeuous two-phase	Perstraction (org. phase)	Reactive perstraction	Supercritical fluids	Ion-exchange	Hydrophobic adsorption	Affinity adsorption	Stripping	Distillation (inc. vacuum)	Transmembrane distillation	Pervaporation	Electrodialysis	Nanofiltrat., rev. osmosis	Precipitation		
fine chemicals	salicylic acid	*Pseudomonas aeruginosa*	b								1									1968	[6]
	7,8-epoxy-1-octene	*Pseudomonas oleovorans*	b	1																1977	[7]
	propeneoxid	*Mycobacterium* sp.	b	2a																1985	[218]
	7,8-epoxy-1-octene	*Pseudomonas putida*	b	1																1986	[219]
	1,2 epoxyalkanes	*Nocardia corallina*	b	1																1986	[220]
	l-tryptophane	*Escherichia coli*	b	1,2a																1987	[221]
	cis-benzeneglycol	*Pseudomonas* sp.	b	1																1987	[222]
	1,2 epoxyoctane	*Nocardia corallina*	b	2a																1990	[223]
	1-tetralol	*Acinetobacter* sp.	b	1																1990	[224]
	hydroxynicotinic acid	*Achromobacter xylosoxidans*	b																1	1991	[225]
	phenylacetylcarbinol	*Saccharomyces cerevisiae*	b	1																1992	[226]
	6-aminocillanic acid	recomb. *Escherichia coli*	b			3														1992	[227]
	2-octanol	*Gluconobacter oxydans*	b				4b													1992	[228]
	phenylacetylcarbinol	*Saccharomyces cerevisiae*	b	1																1992	[229]
	naphthal.-*cis*-glycol	*Pseudomonas putida*	b	2a																1992	[230]
	hydroxyvitamin D3	*Amycolata autorophica*	b								1									1994	[231]
	phenyl glycidyl ether	*Mycobacterium* sp.	b	1																1994	[232]
	styrene epoxide	recomb. *Escherichia coli*	b	1																1994	[233]
	glutamic acid	*Corynebacterium glutamicum*	dn								3									1995	[162]
	2-hydroxy1-phe nyl-1-propanone	*Saccharomyces cerevisiae*	b	2a																1996	[234]
	3-pyridylacetic acid	*Pseudomonas putida*	b							3								4a		1996	[235]
	D-pantoyl lactone	*Fusarium oxysporum*	b	1																1996	[236]
	fluorocatechol	*Pseudomonas putida*	b								3									1997	[237]
	3,4-methylene-dioxyphenyl isopropanol	*Zygosaccharomyces rouxii*	b								4c									1997	[238]
	(*R*)-2-octanol	*Pichia etchellsii*	b	1																1998	[239]
	ethyl 3-oxobutanoate	*Saccharomyces cerevisiae*	b	2a																1998	[240]
	nitroalcohols	*Saccharomyces cerevisiae*	b	1																1998	[241]
	1,2 epoxyhexane	*Rhodotorula glutinis*	b				4a													1999	[242]
	1,2-epoxy-7,8-octene	*Pseudomonas oleovorans*	b				4a													1999	[243]
	N-heterocyclic carboxylic acids	*Pseudomonas putida*	b		1															1999	[244]
	1-octanol	*Pseudomonas oleovorans*	b	1																1999	[35]
	3-phenylcatechol	recomb. *Escherichia coli*	b								3									1999	[245]
steroids	4-androstene-3,17-dione	*Nocardia rhodocrous*	b	2a																1980	[246]
	1-dehydrodiosgenone	*Nocardia rhodocrous*	b								1									1985	[247]
	prednisolone	*Arthrobacter simplex*	b			1														1986	[248]
	11α-hydroxyprogesteron	*Aspergillus ochraceus*	b																	1987	[249]
	4-androstene-3,17-dione	*Mycobacterium fortuitum*	b	2a																1987	[250]
	4-androstene-3,17-dione	*Flavobacterium dehydrogenans*	b	1																1989	[251]
	4-androstene-3,17-dione	*Mycobacterium* sp.	b								1									1989	[252]
	hydrocortisone	*Arthrobacter simplex*	b	1,2a																1990	[253]
	hydrocortisone	*Arthrobacter simplex*	b	2a							2a									1991	[254]
	4-androstene-3,17-dione	*Rhodococcus equi*	b	1																1992	[255]
	hydrocortisone	*Arthrobacter simplex*	b	2a																1992	[256]
	4-androstene-3,17-dione	*Mycobacterium* sp.	b								1									1993	[257]
	11α-hydroxyprogesteron	*Aspergillus ochraceus*	b	2a																1994	[258]
	4-androstene-3,17-dione	*Mycobacterium* sp.	b	2a																1994	[259]
	6β-hydroxyc holest-4-en-3-one	*Pseudomonas* sp.	b	1																1994	[260]
	hydrocortisone	*Arthrobacter simplex*	b	1																1995	[261]
	cholesterone	*Arthrobacter simplex*	b	1																1996	[262]

Table 2 (continued)

Product category	Product	Strain		Extraction						Immobilis.			Evaporation				Perme-ation			Year	Refer-ence
			Denovo (dn), bioconv. (b)	Organic solvent	Reactive	Aqeuous two-phase	Perstraction (org. phase)	Reactive perstraction	Supercritical fluids	Ion-exchange	Hydrophobic adsorption	Affinity adsorption	Stripping	Distillation (inc. vacuum)	Transmembrane distillation	Pervaporation	Electrodialysis	Nanofiltrat., rev. osmosis	Precipitation		
secondary metabo-lites	toxin	*Clostridium tetani*	dn			1														1965	[5]
	cycloheximide	*Streptomyces griseus*	dn								1,2 b									1983	[14]
	red pigment	*Monascus* sp.	dn								1									1984	[263]
	cycloheximide	*Streptomyces griseus*	dn								3										[264]
	longchain unsaturated hydrocarbons	*Botryococcus braunii*	dn	1																	[265]
	thienamycin	*Streptomyces cattleya*	dn									1								1989	[266]
	penicillin G	*Penicillium chrysogenum*	dn	4c																1992	[267]
	cephalosporin C	*Cephalosporium acremo nium*	dn			1														1994	[268]
	pristinamycins	*Streptomyces pristinaespiralis*	dn			1														1994	[269]
	subtilin	*Bacillus subtilis*	dn			1														1994	[270]
	cycloheximide	*Streptomyces griseus*	dn	1																1995	[271]
	gibberellic acid	*Gibberella fujikuroi*	dn	1																1995	[272]
	red pigment	*Serratia* sp.	dn								3									1999	[273]
proteins	cellulase	*Trichoderma reesei*	dn			1														1984	[274]
	amylase	*Bacillus subtilis*	dn			1														1985	[275]
	amylase	*Bacillus subtilis*	dn			1														1987	[276]
	amylase	*Bacillus amyloliquefaciens*	dn			1														1988	[277]
	cellulase	*Trichoderma reesei*	dn			1														1988	[278]
	amylase	*Bacillus amyloliquefaciens*	dn			1														1991	[279]
	cellulase	*Trichoderma reesei*	dn			1														1992	[280]
	amylase	*Bacillus amyloliquefaciens*	dn									4c								1994	[281]
	chitinase	*Serratia marcescens*	dn			1														1994	[282]
	galactosidase	*Escherichia coli*	dn			1														1995	[283]
	heat shock proteins	*Escherichia coli*	dn			1														1996	[284]
	phosphomonoesterase	*Nicotiana tabacum*	dn			1														1996	[285]
	xylanase	*Escherichia coli*	dn			1														1999	[286]

The success of an ISPR process does not depend only on the chosen separation technique but also on the configuration of the bioreactor/separation units and mode of operation. Previous reviews have shown the various possible modes of operation (continuous, batch) and the use of a separation unit outside of the reactor or separation techniques that act right inside the fermenter [19, 22, 31]. Freeman and coworkers introduced a classification scheme for ISPR process based on batch/continuous operation and internal (within the reactor)/external (outside the reactor) removal of the product [3].

However, another criterion that needs to be considered for the selection of a suitable ISPR method is the mode of contact between the microorganisms and the separation phase that removes the product from the vicinity of the cell. Direct contact between the microorganism and a water-immiscible solvent (phase toxicity) or solid adsorbent material can have inhibitory effects on the cell [31, 33]. Therefore, this direct contact limits the choice of separative aids. In addition, stability and robustness of a process is reduced if the cells are in direct

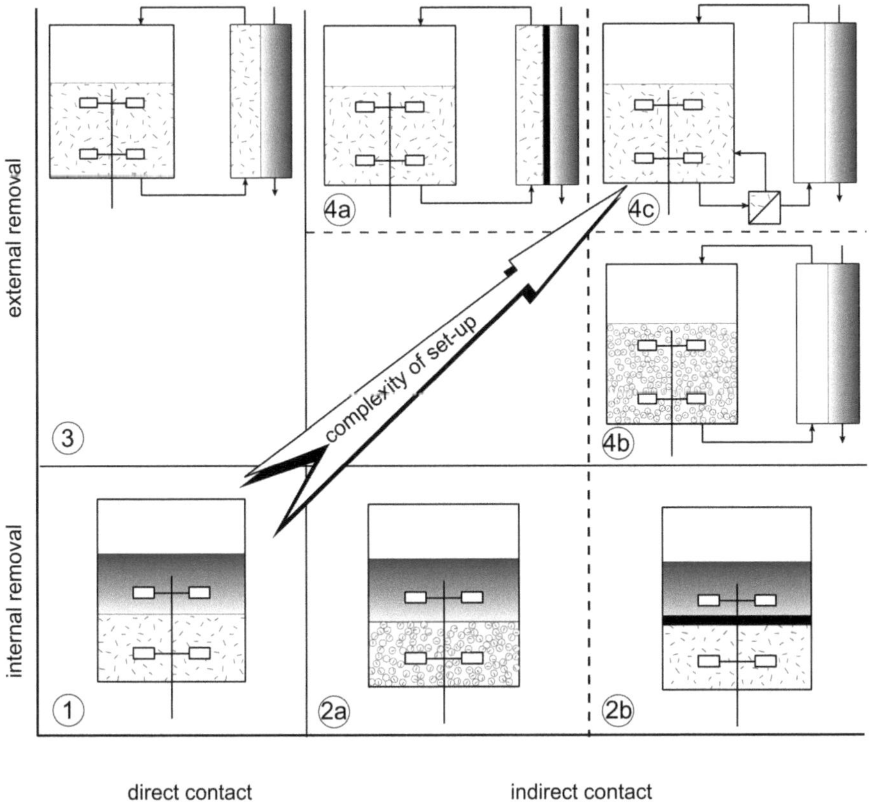

Fig. 1. ISPR Configurations

contact with the separation phase. Stable emulsions are formed if a water-immiscible solvent and a living cell-containing aqueous phase are mixed vigorously. Or, cells can form a biofilm on the adsorbent material, thereby reducing the adsorption capacity of the particles. Therefore, categorization of different ISPR configurations is done in this review by characterizing the position of the in situ separation and the mode of contact between the cell and the separation phase. The in situ removal of the product can take place either inside the reactor (internal) or in an external loop. The contact between the microorganisms and the product separation phase can be either direct or indirect (Fig. 1). The direct contact can take place within the reactor (case 1) or in an external loop (case 3). Direct contact can be prevented within the reactor by immobilizing the microorganisms in a gel matrix such as alginate (case 2a) or by an internal membrane (case 2b). An indirect contact outside the reactor is achieved by three different configurations. The cell-containing reaction medium is circulated in an external loop through membrane modules (case 4a). This set-up is relatively simple; however, fouling or clogging of the membrane in perstraction, pervaporation, or electrodialysis processes has often been observed when the cells get in direct contact with the membranes. Therefore, the reaction medium is often clarified from the cells by either a micro-/ultrafiltration or a centrifugation step before it circulates through an external separation unit (case 4c). Alternatively, a cell-free reaction medium that circulates through an external separation loop is achieved by immobilization of the cells on carriers or in a gel matrix within the reactor (case 4b). A fluidized bed reactor is often used in this configuration.

Cases 2–4 can always be operated either in batch or continuous mode. However, continuous operation is not always possible for case 1. In the case of aqueous two-phase systems or certain aqueous-organic two-phase systems, the separation of the two phases within the reactor needs a certain settling time and therefore allows only semi-continuous operation. Alternatively, it is possible to achieve continuous operation by using an additional external settler. Since the separation still occurs in the reactor, this configuration also belongs to case 1. There is no obligate formation of stable emulsions in anaerobic cultures in the presence of a water-immiscible solvent, since vigorous stirring is not necessary. Consequently, continuous operation is practicable [34, 35]. However, aeration and vigorous stirring in aerobic fermentation form inevitable stable emulsions, which makes a continuous operation impossible.

3.1 Evolution of ISPR Applications

A total of almost 250 ISPR projects in microbial whole cell biotechnology are listed in Table 2. Over one third of these projects have dealt with the production of organic solvents such as ethanol, butanol, acetone or propanol (90 projects). Ethanol (70% of all the solvents) has been by far the most important microbial product for which different ISPR techniques have been applied. The second most important class of products involved in ISPR projects have been organic acids such as lactic, acetic, butyric, or propionic acid (54 projects). Most of effort in this product class has focused on lactic acid (55% of all organic acids). Important ISPR activities have also been reported for the microbial production of various aromas and fine chemicals (30 projects in each product category). A considerable amount of ISPR approaches have been shown in steroid conversions (17 projects) and the production of secondary metabolites and various enzymes (13 projects in each product category).

Various ISPR techniques were investigated to increase the productivity of ethanol production by microbial means. The number of ISPR applications in microbial biotechnology steadily increased in the 1980s (Fig. 2a). It leveled off in the 1990s with about 18 projects per year reported in (patent) literature. However, a closer inspection of the different product categories reveals different tendencies. ISPR activities in the field of microbial production of solvents steadily decrease after having peaked at the end of the 1980s (Fig. 2c). Almost none of the ISPR projects on ethanol production that were tested up to pilot scale ended on an industrial scale, since they were not competitive with the petrochemical industry. In contrast, research effort on ISPR applications for organic acid and aroma production continuously increased in the 1990s (Figs. 2d and e) in concurrence with the raised demand for natural flavors

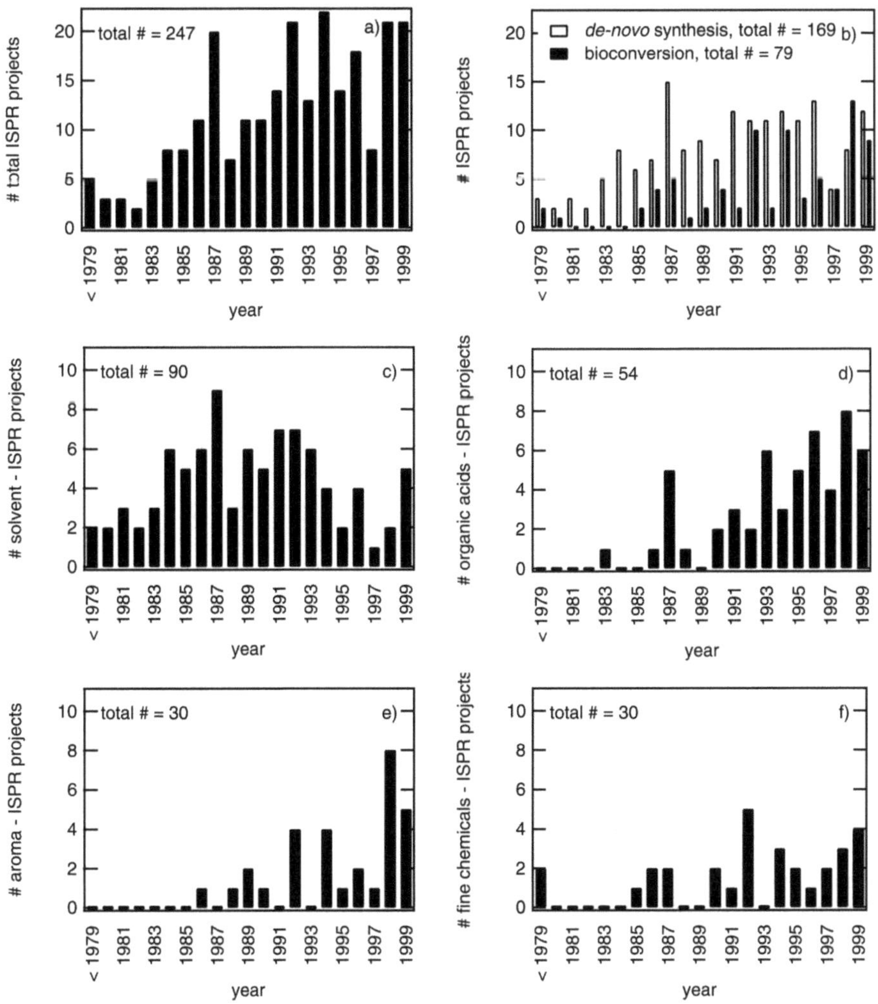

Fig. 2 a–f. Evolution of the number of reported ISPR projects within the last 20 years: **a** total, **b** de novo synthesis and bioconversion, **c** solvents, **d** organic acids, **e** aromas, **f** fine chemicals

and lactic acid (for the production of its biodegradable polymer). The evolution of the ISPR activities reflects well the general market development in microbial biotechnology, in which ethanol production from renewable sources was heavily investigated in the 1980s. Later on, ISPR techniques were also applied to low-volume high-value products, which were increasingly produced by microorganisms.

In general, more and more microbial applications are based on bioconversions, especially in the field of aroma compounds, fine chemicals, and steroid conversions. Therefore, ISPR applications for microbial bioconversions steadily increased in the 1990s (Fig. 2b). Due to the physicochemical similarity of pre-

Table 3. Summary of different reactor/separator configurations for the different ISPR techniques

ISPR configurations		Extraction						Immobilis.			Evaporation				Size		
	Total number of ISPR categories	Organic solvent	Reactive	Aqeuous two-phase	Perstraction (org. Phase)	Reactive perstraction	Supercritical fluids	Ion-exchange	Hydrophobic adsorption	Affinity adsorption	Stripping	Distillation (inc. vacuum)	Transmembrane distillation	Pervaporation	Electrodialysis	Nanofiltrat., rev. osmosis	Preciption
direct contact, internal removal (1)	110	51	4	25				1	15	1	8	3					2
indirect contact, internal removal with immobilized cells (2a)	35	24	3					2	3	1	2						
indirect contact, internal removal (2b)	10	1	1		2	1			3					2			
direct contact, external removal (3)	29	5		2	1			4	11		2	2		1	1		
indirect contact, external removal (4a)	22				5	3							1	9	3	1	
indirect contact, external removal with immobilized cells (4b)	30	3	2		9	2		1	3		2	1		6	1		
indirect contact, external removal with cell recycling (4c)	39	4	6	1	2	1	1	3	6	1		2		3	7	2	
		88	16	28	19	7	1	11	41	3	14	8	1	21	12	3	2
total configuration number of ISPR projects	275	159						55			44				15		2

cursor and product in bioconversions, the application of ISPR is more difficult than for conventional de novo fermentation processes [21].

3.2
Summary of ISPR Techniques and Configurations

Table 3 summarizes the different reported ISPR techniques and assigns them to the different configurations of the reactor/separator set-up. The total number of listed configurations for all the ISPR projects (275) is slightly higher than the total number of ISPR projects (247) listed in this table. This is due to the fact that several different ISPR configurations were included in the same publication.

3.2.1
Techniques

Extraction-based ISPR techniques (159 projects) were used in more than 50% of all the cases (Table 3). Adsorption (55 projects) and evaporation (44 projects)-based systems covered almost 20% each of total ISPR cases. Other ISPR techniques such as electrodialysis or precipitation were used only in a very limited number of cases. Extraction into a water-immiscible organic phase (88 projects) and adsorption onto a hydrophobic resin or carrier (41 projects) are the principal investigated ISPR techniques. Both of them take advantage of the hydrophobic property of the product, which differs from the aqueous background medium. Membrane-assisted ISPR techniques such as pervaporation (21 projects), perstraction into an organic phase (19 projects), and reactive perstraction with ternary and quaternary amines (7 projects) were increasingly investigated in the last ten years. The use of membranes allows continuous product removal and increases either the selectivity (pervaporation) or reduces the toxicity of the extractants towards the cells (perstraction) by avoiding direct contact. Aqueous two-phase systems (28) were primarily applied to the recovery of products that have both hydrophilic and hydrophobic properties such as proteins/enzymes and some secondary metabolites. Stripping (14 projects) and distillation (8 projects) can only be applied to volatile products, such as ethanol, with a boiling point lower than water.

3.2.2
Configurations

Most ISPR projects (110 projects) were done with the simple configuration set-up 1, whereby the inhibitory product is separated from the reaction phase within the fermenter (Table 3). Most of these cases deal with the extraction of the product into an organic (51 projects) or a second aqueous (25 projects) phase. Adsorption (15 projects) on hydrophobic resins and stripping (8 projects) are minor applications of this configuration. This high total number of ISPR projects using configuration 1 was also achieved because many of these publications reported only batch experiments in shake flasks. Set-up 3, the other configuration allowing a direct contact between the cell and the separative driving force, was predominantly used with hydrophobic resins, which were placed in an external column. Thereby, problems with the abrasive impact of the resins within the reactor vessel were avoided, and on-line regeneration of the external column was possible. Immobilization of the cells in a gel matrix protects the cells from the separative force within the reactor (configuration 2a). This set-up was mainly used in the presence of an organic phase to prevent its direct contact with the cells (24 projects). Configuration 2b, membrane system inside the reactor vessel, was not used very frequently (10 projects) because such a system is not commercially available and suitable to scale-up.

One third (91 projects) of all the reported ISPR projects in microbial biotechnology belong to category 4, in which the separation of the product takes place outside of the reactor and the cells are not in direct contact with the separative force. This configuration allows continuous processing and reduces the interaction between the cells and the separating device, which leads to a robust system with a long-term operability. However, additional equipment and control units add a higher complexity to the system, which is probably only applicable for the production of a dedicated single product and not a multipurpose plant. There were a total of 22 projects using external membrane modules for pervaporation transmembrane distillation, perstraction, and electrodialysis (configuration 4a), which operate with unclarified reaction suspensions. The clarification of the reaction suspension from the cells was equally done either by immobilization of the cells (configuration 4b, 30 projects) or the introduction of a micro-/ultrafiltration device or a centrifuge (configuration 4c, 39 projects). Interestingly enough there were more reported projects of configurations 4b and 4c using external membrane modules, which consisted of an additional clarifying step (configuration 4b and 4c for membrane processes, total of 33 projects), than there were projects that used external membrane modules without an additional clarifying step (configuration 4a, 22 projects). This additional step was introduced to prevent the membrane operation from clogging and fouling.

4
Evaluation of an Appropriate ISPR Technique

Almost 250 ISPR projects are mentioned in the previous sections that were developed according to the physicochemical properties of their product. Many also

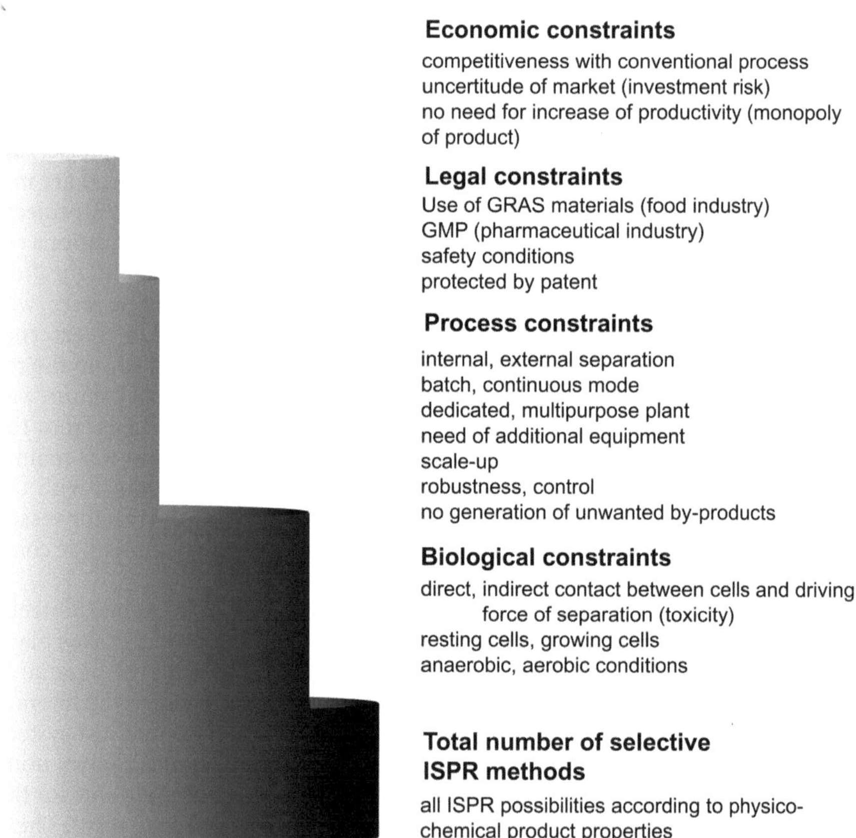

Fig. 3. Evaluation of an appropriate ISPR technique

reported an increase of productivity [3]. Most of the work was done on a laboratory scale and some on a pilot scale. Mainly ISPR processes for microbial ethanol production were transferred to pilot scale. Detailed economic assessments for these different processes were published [36, 37]; however, only the BIOSTIL process finally succeeded on an industrial scale, primarily due to its use of conventional and reliable equipment. This process was mainly introduced in Brazil, where sugarcane is excessively available and inexpensive as a renewable raw material. Citric and lactic acid production with the coprecipitation of their calcium salts was also introduced on industrial scale.

Cargill Dow Polymers LLC (Midland, USA) is presently constructing a polylactide production facility in Blair (Nebraska, USA) that will be completed in 2002 and will produce 140,000 t $year^{-1}$ of the biodegradable natural polymer. The company is supposed to transfer their patented ISPR process for extractive lactic acid fermentation on an industrial scale [38]. Pressurized CO_2 (17 bar) is used as an organic phase to extract lactic acid from the prior clarified reaction suspension. Tertiary amines support the extraction as they form stable ion pairs with the undissociated carboxylic acid, which results in a much higher extraction efficiency [39]. Carbon dioxide is also used as an acidifying agent during the extraction of lactic acid, and the formed carbonates are recycled to the fermentation and are used as base for pH control. This results in a process in which the consumption of acids and bases is avoided and in which the generation of waste salts is eliminated. Lactic acid is recovered by a back-extraction from the organic phase, which is continuously recycled to the extractor.

It is obvious that eventually the economic considerations of a chosen ISPR process is the decisive factor. Constraints other than the physicochemical properties of the target product need to be considered early on for an economically viable process; Figure 3 gives an overview of these. The influence of the listed points on the overall economics also needs to be evaluated as early as possible. As mentioned above, the decision for an internal/external separation and a direct/indirect contact between the cells and the separative force is essential for the success of an ISPR process. The most important constraints that definitely need to be considered are discussed below.

4.1 Biological Constraints

- The type of microbial strain is decisive for the choice of contact between cell and separative force. *Pseudomonas* species for instance, support direct contact with more polar solvents than *Saccharomyces cerevisiae* [40]. A correlation between the solvent toxicity and its hydrophobicity was obtained by plotting the cellular activity retention against the Hansch parameter $\log P_{oct}$, which gives the logarithm of the partition coefficient of the solvent in the octanol-water two-phase system [30, 76].
- Different impacts on the choice of an ISPR process arise from the use of growing or resting cells. Besides introducing additional mass transport limitation, cell immobilization in beads is only practical for processes with resting cells due to limited space in the gel matrix. In addition, growing cells generally sup-

port less harsh environmental conditions than resting cells. The choice of solvent for an organic two-phase system is more restricted for growing than for resting cells.
- Strictly aerobic conditions usually reduce the degree of freedom of an ISPR set-up. The oxygen demand of the cells in an external loop for instance can cause problems. Other limitations in an aerobic system can arise through vigorous stirring. This causes difficulties in the application of an organic phase (formation of stable emulsions) or hydrophobic carriers (abrasion) in the reactor vessel.
- Bioreactor by-passes not only cause problems of oxygen limitation but also other substrate or product gradients. A nutrient limitation, for instance, can lead to the sporulation of various *Bacillus* species. In addition, the appearance of increased shear stress in the external loop can cause problems.

4.2
Process Constraints

The mode of operation (batch, continuous) and the use of a dedicated or a multipurpose plant is crucial for the choice of an appropriate ISPR configuration. The successful production of bulk products such as lactic acid requires an optimized dedicated system. This production plant has special equipment and advanced control strategies and is not very flexible. Generation of additional by-products has to be avoided. High added value, low volume products such as fine chemical or natural flavors are mostly produced in multipurpose plants. Investment in additional equipment or modifications of the reaction vessels is more difficult to justify, if just a few products need the application of ISPR. In this case, the less complex ISPR configurations 1 and 3 are preferred. Furthermore, the outlet of the production should also be easily adjustable to the market demand.

4.3
Legal Constraints

- It always needs to be checked if the extension of a manufacturing facility with an ISPR system still allows production under the required regulatory norms. Compliance with GMP regulations is indispensable for pharmaceutical products and additional material for the production of food additives need to GRAS approved.
- Extension of the production facility with an ISPR system may introduce a safety problem. The use of hazardous or flammable solvents for an extraction system for instance needs additional safety precautions.
- An appropriate ISPR extension may be protected by a patent.

4.4
Economical Constraints

Many investigated ISPR processes were not introduced on an industrial scale because they were not competitive with traditional technology. In addition, the uncertainty of the market demand was often too risky for such an investment. On the other hand, there is sometimes no need for an increase of the productivity, especially for high value products, since the company is the sole manufacturer of the product.

5 Future Trends of ISPR in Biotechnology

The excellent review by Freeman and coworkers predicted a shift of the application of ISPR techniques from bulk products to high added value, low volume products such as fine chemicals, food additives, or high molecular products [3]. This shift has occurred to a certain extent; however, as seen beforehand there is still a lot of research going on for bulk products such as lactic acid. It seems that there is still a need for dedicated ISPR facilities, that use robust equipment and that do not generate a lot of unwanted by-products. A reason for the failure of many previous ISPR projects was the addition of an extension that was too complex. Another requirement for a successful ISPR introduction is certainly the use of a flexible and broadly applicable multipurpose plant. It should be possible for instance to enhance the production of several different flavors with the same ISPR configuration, or at least with minor modifications. Thus, a compromise is needed for the choice of separation (e.g., choice in the type of hydrophobic resin, solvent, or membrane) between high selectivity for one single product or a less selective, but broad applicable solution.

There is still a lack of highly selective separation techniques with a high capacity, especially for the ISPR application of high added value products. A more frequent use of different affinity recognition-based separation techniques for ISPR applications was predicted by several authors [1, 3, 19]. There is a lot of research going on in the broad field of affinity separation in downstream processing; however, results have been rarely coupled in situ to fermentation processes. The successful use of adsorptive membranes was demonstrated for the in situ recovery of a tissue plasminogen activator produced by recombinant animal cells [41]. This method for the purification of biomolecules by a combination of affinity interactions for the target molecule and membrane filtration for unwanted material has found interest in downstream processing [42–44]. Conceptually, membrane-based affinity purification systems enable high volumetric throughputs while rejecting the cells. Another affinity-based ISPR application is the use of molecular imprinted polymer (MIP) adsorbents. This has been successfully demonstrated for the in situ recovery of a decalactone flavor produced by fungi [45]. Molecular imprinting is an emerging technique in which polymeric adsorbents are synthesized that exhibit highly selective binding for a particular molecule [46–48]. This technique is widely tested in conventional sequential downstream operations. However, their applications in ISPR applications are still very rare. The major drawback is currently their low capacity for the target molecule.

Another promising downstream technique that could be used for ISPR systems is counter-current chromatography [1]. This technique uses two immiscible phases to separate solutes on the basis of their relative solubility in the two solvents. It is essentially an intensive liquid-liquid extraction process and allows

Table 4. Selected ISPR projects with plant and animal cells

Product category	Product	Cell species		Extraction						Immobilis.			Evaporation				Size			Year	Reference
			Denovo, bioconversion	Organic solvent	Reactive	Aqeuous two-phase	Perstraction (org. phase)	reactive perstraction	Supercritical fluids	Ion-exchange	Hydrophobic adsorption	Affinity adsorption	Stripping	Distillation (inc. vacuum)	Transmembrane distillation	Pervaporation	Electrodialysis	Nanofiltrat., rev. osmosis	Precipitation		
plant cells																					
aroma	vanillin, vanillic acid	*Capsicum frutescens*	b								1									1999	[287]
secondary	anthraquinones	*Cinchona ledgeriana*	dn								1									1986	[288]
metabo	indole alkaloids	*Catharanthus roseus*	dn								1									1988	[289]
lites	ajmalicine	*Catharanthus roseus*	dn								1									1989	[290]
	thiophenes	*Tagetes patula*	dn			1														1992	[291]
	solavetivone	*Hyoscyamus muticus*	dn	3							3									1993	[292]
	caffeine	*Coffea arabica*	dn								1									1994	[293]
	benzophenanthri dine alkaloids	*Escherichia californica*	dn	1																1994	[294]
	anthraquinones	*Morinda citrifolia*	dn	1																1996	[295]
	taxol	*Taxus brevifolia*	dn	1																1996	[296]
	digoxin	*Digitalis lanata*	b								1									1998	[297]
animal cells																					
proteins	tissue plasminogen activator	human fibroblasts	dn								4b									1991	[298]
	monoclonal antibody	hybridoma cells	dn			1														1996	[299]
	tissue plasminogen activator	hybridomacells	dn									4c								1998	[41]
by-product	ammonium	hybridoma cells	dn							4c										1991	[300]
	ammonium	hybridoma cells	dn													2b 4a				1994	[301]
	ammonium, lactate	hybridomacells	dn														4c			1995	[302]

chromatographic-quality separations but with a much greater capacity than conventional solid adsorbents. However, successful in situ applications of this technique to fermentation processes have not yet been reported.

Direct product precipitation allows pure recovery without need of further separation steps. Contamination by organic solvents is thereby avoided, which is often a problem for drug and food applications. A fermentation process that removes lactic acid by in situ crystallization with calcium ions was one of the first successful applications of a whole cell ISPR process [11]. Recently, the potential of in situ precipitation was shown for the solid-solid enzymatic conversion of Ca-maleate to Ca-D-malate. In situ crystallization processes employing whole cells are expected to be increasingly applied in the recovery of carboxylic acids, antibiotics, and proteins.

ISPR applications have also been reported with raising interest for plant and animal cell technology. Table 4 does not give a complete summary on the ISPR activities with these two types of cells, but it gives a good overview for their main applications. Plant cells are mostly used for the production of secondary metabolites. Besides the technique of permeabilization of the cell wall and elicitation of the product formation, ISPR has proven success for plant cell cultures on a laboratory scale [49]. Animal cells are widely used for the production of various recombinant proteins. One successful ISPR application is the removal of toxic by-products such as ammonium or lactate. On the other hand, different techniques have been tested to reduce the number of downstream steps by recovering the target protein in situ from the cell suspension (Table 4).

Acknowledgement. The authors gratefully acknowledge help by André Jaquet in preparing this review.

6
References

1. Lye GJ, Woodley JM (1999) TIBTECH 17:395
2. Holst O, Mattiasson B (1991) In: Mattiasson B, Holst O (eds) Extractive bioconversions, vol 11. Marcel Dekker, p 11
3. Freeman A, Woodley JM, Lilly MD (1993) Bio/Technology 11:1007
4. Park CH, Geng QH (1992) Sep Purific Meth 21:127
5. Puziss M, Heden CG (1965) Biotechnol Bioeng 7:355
6. Kitai A, Tone H, Ishikura T, Ozaki A (1968) J Ferment Technol 46:442
7. Schwartz RD, McCoy CJ (1977) Appl Environ Microbiol 34:47
8. Cysewski GR, Wilke CR (1977) Biotechnol Bioeng 19:1125
9. Friedman MR, Gaden EL (1970) Biotechnol Bioeng 12:961
10. Kominek LA (1975) Antimicrob Agents Chemother 7:861
11. Buchta K (1983) In: Dellweg H (ed) Biotechnology, vol 3, p 410
12. Ehnström L, Frisenfelt J, Danielsson M (1991) In: Mattiasson B, Holst O (eds) Extractive bioconversions. Marcel Dekker, p 303
13. Alfa-Laval (1981) Tech Rept IB-8104-03, Tumba, Sweden
14. Wang HY (1983) Ann N Y Acad Sci 413:313
15. Roffler SR, Blanch HW, Wilke CR (1984) TIBTECH 2:129
16. Daugulis AJ (1988) Biotechnol Prog 4:113
17. Kobayashi T (1990) Ann N Y Acad Sci 22:248
18. Mattiasson B, Holst O (1991) Extractive bioconversions. Marcel Dekker
19. Schügerl K, Kretzmer G, Freitag R, Scheper T (1994) Chem Ing Tech 66:1585
20. Daugulis AJ (1997) Curr Opin Biotechnol 8:169
21. Chauhan RP, Woodley JM (1997) CHEMTECH 27:26
22. Groot WJ, van der Lans RGJM, Luyben KCAM (1992) Process Biochem 27:61
23. Durre P (1998) Appl Microbiol Biotechnol 49:639
24. Andersson E, Hahn-Hägerdal B (1990) Enzyme Microb Technol 12:242
25. Zijlstra GM, Degooijer CD, Tramper J (1998) Curr Opin Biotechnol 9:171
26. Diaz M (1988) TIBTECH 6:126
27. Woodley JM, Harrop AJ, Lilly MD (1990) Ann N Y Acad Sci 613:191-200
28. Tramper J, Vermue MH (1993) Chimia 47:110
29. Nikolova P, Ward OP (1993) J Ind Microbiol 12:76
30. Leon R, Fernandes P, Pinheiro HM, Cabral JMS (1998) Enzyme Microb Technol 23:483
31. Holst O, Mattiasson B (1991) In: Mattiasson B, Holst O (eds) Extractive bioconversions, vol 11. Marcel Dekker, p 189
32. Pörtner R, Märkl H (1998) Appl Microbiol Biotechnol 50:403
33. Bruce LJ, Daugulis AJ (1991) Biotechnol Prog 7:116
34. Zigova J, Sturdik E, Vandak D, Schlosser S (1999) Process Biochem 34:835
35. Mathys RG, Schmid A, Witholt B (1999) Biotechnol Bioeng 64:459
36. Daugulis AJ, Axford DB, McLellan PJ (1991) Can J Chem Eng 69:488

37. Sundquist J, Blanch HW, Wilke CR (1991) In: Mattiasson B, Holst O (eds) Extractive bioconversion. Dekker, p 237
38. Baniel AM, Eyal AM, Mizrahi J, Hazan B, Fisher RR, Kolstad JJ, Stewart BF (1996) US Patent 5 510 526
39. van Halsema FED, van der Wielen LAM, Luyben KCAM (1998) Ind Eng Chem Res 37:748
40. Inoue A, Horikoshi K (1991) J Ferment Bioeng 71:194
41. Vogel JH, Kroner KH, Anspach FB (1998) Chem Ing Tech 70:1151
42. Ling TGI, Mattiasson B (1989) Biotechnol Bioeng 34:1321
43. Nachman M, Azad ARM, Bailon P (1992) Biotechnol Bioeng 40:564
44. Agrawal A, Burns MA (1996) Biotechnol Bioeng 52:539
45. Dukler A, Freeman A (1998) J Mol Recognit 11:231
46. Mosbach K, Ramström O (1996) Bio/Technology 14:163
47. Ju JY, Shin CS, Whitcombe MJ, Vulfson EN (1999) Biotechnol Bioeng 64:232
48. Takeuchi T, Haginaka J (1999) J Chromat B 728:1
49. Dornenburg H, Knorr D (1995) Enzyme Microb Technol 17:674
50. Ramalingham A, Finn RK (1977) Biotechnol Bioeng 19:583
51. Maiorella B, Wilke CR (1980) Biotechnol Bioeng 22:1749
52. Kühn I (1980) Biotechnol Bioeng 22:2393
53. Lee JH, Woodard JC, Pagan RJ, Rogers PL (1981) Biotechnol Lett:177
54. Hahn-Hägerdal B, Mattiasson B, Albertsson PA (1981) Biotechnol Lett 3:53
55. Wang HY, Robinson FM, Lee SS (1981) Biotechnol Bioeng Symp 11:555
56. Lee SS, Wang HY (1982) Biotechnol Bioeng Symp 12:221
57. Minier M, Goma G (1982) Biotechnol Bioeng 24:1565
58. Inloes DS, Taylor DP, Cohen SN, Michaels AS, Robertson CR (1983) Appl Environ Microb 46:264
59. Lencki RW, Robinson CW, Moo-Young M (1983) Biotechnol Bioeng Symp 13:617
60. Walsh PK, Liu CP, Findley ME, Liapsis AI, Siehr DJ (1983) Biotechnol Bioeng Symp 13:629
61. Larsson M, Holst O, Mattiasson B (1984) Third European Congress on Biotechnology, vol 2, p 10
62. Groot WJ, Schoutens GH, van Beelen PN, van den Oever CE, Kossen NWF (1984) Biotechnol Lett 6:789
63. Matsumura M, Märkl H (1984) Appl Microbiol Biotechnol 20:371
64. Larsson M, Mattiasson B (1984) Ann N Y Acad Sci 434:144
65. Andrews GF, Elkcechens S (1984) Chem Eng Commun 29:139
66. Ghose TK, Roychoudhury PK, Ghosh P (1984) Biotechnol Bioeng 26:377
67. Taya M, Ishii S, Kobayashi T (1985) J Ferment Technol 63:181
68. Dale MC, Okos MR, Wankat PC (1985) Biotechnol Bioeng 27:932
69. Chung IS, Lee YY (1985) Enzyme Microb Technol 7:217
70. Sato K, Nakamura K, Sato S (1985) Biotechnol Bioeng 27:1312
71. Groot WJ, Luyben KCAM (1986) Appl Microbiol Biotechnol 25:29
72. Garcia A, Iannotti EL, Fischer JL (1986) Biotechnol Bioeng 18:785
73. Honda H, Taya M, Kobayashi T (1986) J Chem Eng Jpn 19:268
74. Kobayashi T (1986) Japanese Patent 61192291
75. Cho T, Shuler ML (1986) Biotechnol Prog 2:53
76. Matsumura M, Märkl H (1986) Biotechnol Bioeng 28:534
77. Choudhury JP, Ghosh P (1986) Biotechnol Lett 8:731
78. Wayman M, Parekh R (1987) J Ferment Technol 65:295
79. Eckert G, Schügerl K (1987) Appl Microbiol Biotechnol 27:221
80. Ennis BM, Qureshi N, Maddox IS (1987) Enzyme Microb Technol 9:672
81. Larrayoz MA, Puigjaner L (1987) Biotechnol Bioeng 30:692
82. Dale MC, Wankat PC, Okos MR (1987) US Patent 4 665 027
83. Roffler SR, Blanch HW, Wilke CR (1987) Bioprocess Eng 2:1
84. Tanaka H, Harada S, Kurosawa H, Yajima M (1987) Biotechnol Bioeng 30:22
85. Nakao SI, Saitoh F, Asakura T, Toda K, Kimura S (1987) J Membrane Sci 30:273
86. Minami T (1987) Japanese Patent 62022593

87. Bar R (1988) J Chem Tech Biotechnol 43:49
88. Roffler SR, Wilke CR, Blanch HW (1988) Biotechnol Bioeng 32:192
89. Jeon YJ, Lee YY (1989) Enzyme Microb Technol 11:575
90. Eiterman MA, Gainer JL (1989) Appl Microbiol Biotechnol 30:614
91. Shukla R, Kang W, Sirkar KK (1989) Biotechnol Bioeng 34:1158
92. Groot WJ, van der Lans RGJM, Luyben KCAM (1989) Appl Microbiol Biotechnol 32:305
93. Calibo RL, Matsumura M, Kataoka H (1989) J Ferment Bioeng 67:40
94. Udriot H, Ampuero S, Marison IW, von Stockar U (1989) Biotechnol Lett 11:509
95. Kollerup F, Daugulis AJ (1989) US Patent 4 865 973
96. Cho T, Shuler ML (1990) US Patent 4 940 547
97. Mori Y, Inaba T (1990) Biotechnol Bioeng 36:849
98. Kang W, Shukla R, Sirkar KK (1990) Biotechnol Bioeng 36:826
99. Christen P, Minier M, Renon H (1990) Biotechnol Bioeng 36:116
100. Bandini S, Gostoli C (1990) In: Pyle DL (ed) Separations for biotechnology 2. Elsevier Applied Science, London, p 539
101. Park CH, Okos MR, Wankat PC (1991) Biotechnol Prog 7:185
102. Friedl A, Qureshi N, Maddox IS (1991) Biotechnol Bioeng 38:518
103. Oliveira AC, Cabral JMS (1991) J Chem Tech Biotechnol 52:219
104. Shabtai Y, Chaimovitz S, Freeman A, Katchalski-Katzir E, Linder C, Nemas M, Perry M, Kedem O (1991) Biotechnol Bioeng 38:869
105. Tedder DW (1991) US Patent 5 036 005
106. Matsumura M, Takehara S, Kataoka H (1992) Biotechnol Bioeng 39:148
107. Qureshi N, Maddox IS (1992) Appl Biochem Biotechnol 34/35:441
108. Kim YJ, Weigand WA (1992) Appl Biochem Biotechnol 34:419
109. Groot WJ, Kraayenbrink MR, Waldram RH, van der Lans RGJM, Luyben KCAM (1992) Bioprocess Eng 8:99
110. Turpin JL, Eldridge RB, Booth-McGee S (1992) US Patent 5 110 319
111. Weilnhammer C, Blass E (1992) In: Kreysa G, Driesel AJ (eds) 10th DECHEMA Annual Meeting of Biotechnologists (DECHEMA biotechnology conferences). VCH, Karlsruhe, p 197
112. Bruce LJ, Daugulis AJ (1992) Biotechnol Lett 14:71
113. Chang HN, Yang JW, Park YS, Kim DJ, Han KC (1992) J Biotechnol 24:329
114. Scott CD (1993) US Patent 5 270 189
115. Davison BH, Thompson JE (1993) Appl Microbiol Biotechnol 39:415
116. Araki K, Minami T, Sueki M, Kimura T (1993) J Soc Ferment Bioeng 41:1007
117. Mollah AH, Stuckey DC (1993) Enzyme Microb Technol 15:200
118. Grobben NG, Eggink G, Cuperus FP, Huizing HJ (1993) Appl Microbiol Biotechnol 39:494
119. Jones TD, Havard JM, Daugulis AJ (1993) Biotechnol Lett 15:871
120. Geng QH, Park CH (1994) Biotechnol Bioeng 43:978
121. Shah MM, Lee YY (1994) Appl Biochem Biotechnol 45–6:585
122. Daugulis AJ, Axford DB, Ciszek B, Malinowski JJ (1994) Biotechnol Lett 16:637
123. Weilnhammer C, Blass E (1994) Chem Eng Technol 17:365
124. Qureshi N, Maddox IS (1995) J Ferment Bioeng 80:185
125. Maddox IS, Qureshi N, Robertsthomson K (1995) Process Biochem 30:209
126. Gapes JR, Nimcevic D, Friedl A (1996) Appl Envir Microbiol 62:3210
127. O'Brien DJ, Craig JC (1996) Appl Microbiol Biotechnol 44:699
128. Gyamerah M, Glover J (1996) J Chem Tech Biotechnol 66:145
129. Moritz JW, Duff SJB (1996) Biotechnol Bioeng 49:504
130. Oliveira AC, Rosa MF, Cabral JMS, Aires-Barros MR (1997) Bioprocess Eng 16:349
131. Kishimoto M, Nitta Y, Kamoshita Y, Suzuki T, Suga K (1997) J Ferment Bioeng 84:449
132. Taylor F, Kurantz MJ, Goldberg N, Craig JC (1998) Biotechnol Lett 20:67
133. Oliveira AC, Rosa MF, Cabral JMS, Airesbarros MR (1998) J Molec Cat B: enzymatic 5:29
134. Qureshi N, Blaschek HP (1999) Biotechnol Prog 15:594
135. Ishizaki A, Michiwaki S, Crabbe E, Kobayashi G, Sonomoto K, Yoshino S (1999) J Biosci Bioeng 87:352

136. Kapucu H, Mehmetoglu U (1999) Appl Microbiol Biotechnol 75:205
137. Dasilva FLH, Rodrigues MI, Maugeri F (1999) J Chem Tech Biotechnol 74:176
138. Domínguez JM, Cao N, Gong CS, Tsao GT (2000) Biotechnol Bioeng 67:336
139. Hongo M, Nomura Y, Iwahara M (1986) Appl Environ Microb 52:314
140. Nuchnoi P, Izawa I, Nishio N, Nagai S (1987) J Ferment Technol 65:699
141. Boyaval P, Corre C, Terre S (1987) Biotechnol Lett 9:207
142. Nomura Y, Iwahara M, Hongo M (1987) Biotechnol Bioeng 30:788
143. Czytko M, Ishii K, Kawai K (1987) Chem Ing Tech 59:952
144. Bar R, Gainer JL (1987) Biotechnol Prog 3:109
145. Nielsen L, Larsson M, Holst O, Mattiasson B (1988) Appl Microbiol Biotechnol 28:335
146. Evans PJ, Wang HY (1990) Appl Microbiol Biotechnol 32:393
147. Yao P, Toda K (1990) J Gen Appl Microbiol 36:111
148. Yabannavar VM, Wang DIC (1991) Biotechnol Bioeng 37:1095
149. Nomura Y, Yamamoto K, Ishizaki A (1991) J Ferment Bioeng 71:450
150. Seevaratnam S, Holst O, Mattiasson B (1991) Bioprocess Eng 6:35
151. Srivastava A, Roychoudhury PK, Sahai V (1992) Biotechnol Bioeng 39:607
152. Lewis VP, Yang S-T (1992) Biotechnol Prog 8:104
153. von Eysmondt J, Vasic-Racki D, Wandrey C (1993) Chem Biochem Eng Q 7:139
154. Vaccari G, Gonzalez-Vara A, Campi AL, Dosi E, Brigidi P, Matteuzzi D (1993) Appl Microbiol Biotechnol 40:23
155. Favre-Bulle O, Weenink E, Vos T, Preusting H, Witholt B (1993) Biotechnol Bioeng 41:263
156. Boyaval P, Seta J, Gavach C (1993) Enzyme Microb Technol 15:683
157. Zhang S, Matsuoka H, Toda K (1993) J Ferment Bioeng 75:276
158. Wang J, Liu P, Zhou D (1994) Biotechnol Tech 8:905
159. Vonktaveesuk P, Tonokawa M, Ishizaki A (1994) J Ferment Bioeng 77:508
160. Dissing U, Mattiasson B (1994) Biotechnol Lett 16:333
161. Zhang S-T, Toda K (1994) J Ferment Bioeng 77:288
162. Einicke WD, Mauersberger P, Handel J, Breuel A (1995) Chem Technik 47:261
163. Katzbauer B, Cesi V, Narodoslawsky M, Moser A (1995) Chem Biochem Eng 9:79
164. Honda H, Toyama Y, Takahashi H, Nakazeko T, Kobayashi T (1995) J Ferment Bioeng 79:589
165. Zihao W, Kefeng Z (1995) Biotechnol Bioeng 47:1
166. Solichien MS, Obrien D, Hammond EG, Glatz CE (1995) Enzyme Microb Technol 17:23
167. Cao N, Du J, Gong CS, Tsao GT (1996) Appl Environ Microb 62:2926
168. Zheng YZ, Ding XH, Yang CW, Cen PL, Tsao GT (1996) Appl Microbiol Biotechnol 57–8:627
169. Kwon YJ, Kaul R, Mattiasson B (1996) Biotechnol Bioeng 50:280
170. Planas J, Radstrom P, Tjerneld F, Hahnhagerdal B (1996) Appl Microbiol Biotechnol 45:737
171. Ye KM, Jin S, Shimizu K (1996) J Ferment Bioeng 81:240
172. Yang ST (1996) US Patent 5 563 069
173. Gaddy JL (1997) US Patent 5 593 886
174. Chan EC, Kuo J (1997) Enzyme Microb Technol 20:585
175. Roddick FA, Britz ML (1997) J Chem Tech Biotechnol 69:383
176. Chen RF, Lee YY (1997) Appl Microbiol Biotechnol 63–5:435
177. Wieczorek S, Brauer H (1998) Bioprocess Eng 18:75
178. Chen CC, Ju LK (1998) Sep Sci Technol 33:1423
179. Matsumoto M, Yuba S, Kondo K (1998) J Chem Eng Jap 31:996
180. Tong Y, Hirata M, Takanashi H, Hano T, Matsumoto M, Shigenobu M (1998) Sep Sci Technol 33:1439
181. Tsao GT, Lee SJ, Tsai G-J, Seo J-H, McQuigg DW, Vorhies SL, Iyer G (1998) US 5 786 185
182. Kollmer A, Rudolf von Rohr P (1998) Chem Ing Tech 70:1125
183. Rothen SA, Sauer M, Sonnleitner B, Witholt B (1998) Biotechnol Bioeng 58:356
184. Jin ZW, Yang ST (1998) Biotechnol Prog 14:457

185. Sun Y, Li YL, Bai S, Hu ZD (1999) Ind Eng Chem Res 38:3290
186. Iyer PV, Lee YY (1999) Appl Biochem Biotechnol 77–79:409
187. Gu Z, Rickert DA, Glatz BA, Glatz CE (1999) LAIT 79:137
188. González-Vara A, Vaccari G, Dosi E, Trilli A, Rossi M, Matteuzzi D (2000) Biotechnol Bioeng 67:147
189. Brookes IK, Lilly MD (1986) Enzyme Microb Technol 8:53
190. Klingenberg A, Hanssen H-P (1988) Chem Biochem Eng Q 2:222
191. de Wulf O, Thonart P (1989) Appl Biochem Biotechnol 20/21:165
192. Kühne B, Sprecher E (1989) Flavour Fragrance J 4:77
193. Berger RG, Drawert F (1990) In: Spiess WEL, Schuberts M (eds) Proceedings of the International Congress. Elsevier, London, p 455
194. Creuly C, Larroche C, Gros JB (1992) Enzyme Microb Technol 14:669
195. Nikolova P, Ward OP (1992) In: Tramper J, Vermue MH, Beeftink HH, von Stockar U (eds) Biocatalysis in non-conventional media. Elsevier, Amsterdam, p 667
196. van der Schaft PH, ter Burg N, van den Bosch S, Cohen AM (1992) Appl Microbiol Biotechnol 36:712
197. Bengtson G, Böddeker KW (1992) Biotechnol Technol 6:23
198. Kawakami K, Nakahara T (1994) Biotechnol Bioeng 43:918
199. Kogan N, Freeman A (1994) Process Biochem 29:671
200. Morin A, Raymond Y, Cormier F (1994) Proc Biochem 29:437
201. Wu W, Sidhoum M, Delancey GB (1994) Biotechnol Bioeng 44:1217
202. Krings U, Berger RG (1995) Biotechnol Tech 9:19
203. Oda S, Inada Y, Kato A, Matsudomi N, Ohta H (1996) J Ferment Bioeng 80:559
204. Lamer T, Spinnler HE, Souchon I, Voilley A (1996) Process Biochem 31:533
205. Molinari F, Aragozzini F, Cabral JMS, Prazeres DMF (1997) Enzyme Microb Technol 20:604
206. Lee IY, Volm TG, Rosazza JPN (1998) Enzyme Microb Technol 23:261
207. Simmonds J, Robinson GK (1998) Appl Microbiol Biotechnol 50:353
208. Doig SD, Boam AT, Leak DI, Livingston AG, Stuckey DC (1998) Biotechnol Bioeng 58:587
209. Doig SD, Boam AT, Leak DJ, Livingston AG, Stuckey DC (1998) Biocat Biotransform 16:27
210. Souchon I, Spinnler HE, Dufossé L, Voilley A (1998) Biotechnol Tech 12:109
211. Molinari F, Villa R, Aragozzini F (1998) Biotechnol Lett 20:41
212. Stentelaire C, Lesage-Meessen L, Delattre M, Haon M, Sigoillot JC, Colonna Ceccaldi B, Asther M (1998) World J Microbiol Biotechnol 14:285
213. Fabre CE, Condoret JS, Marty A (1999) Biotechnol Bioeng 64:392
214. Dufosse L, Souchon I, Feron G, Latrasse A, Spinnler HE (1999) Biotechnol Prog 15:135
215. Souchon I, Dufossé L, Féron G, Voilley A, Spinnler HE (1999) 5th Weurman Flavour Research Symposium, Garching, Germany
216. Schäfer T, Bengtson G, Pingel H, Böddeker KW, Crespo JPSG (1999) Biotechnol Bioeng 62:412
217. Molinari F, Gandolfia R, Aragozzinia F, Leonb R, Prazeresb DMF (1999) Enzyme Microb Technol 25:729
218. Brink LES, Tramper J (1985) IFAC modelling and control of biotechnological processes. Noordwijkerhout, p 111
219. Harbron S, Smith BW, Lilly MD (1986) Enzyme Microb Technol 8:85
220. Furuhashi K, Shintani M, Takagi M (1986) Appl Microbiol Biotechnol 23:218
221. Ribeiro MH, Cabral JMS, Fonseca (1987) Biocatalysis in organic media. Elsevier, Wageningen, p 323
222. van den Tweel WJJ, Marsman EH, Vorage MJAW, Tramper J, de Bont JAM (1987) International Conference on Bioreactors and Biotransformations (Bioreactors and Biotransformations). Elsevier Applied Science Publishers, Gleneagles, Scotland, p 231
223. Kawakami K, Tsuruda S, Miyagi K (1990) Biotechnol Prog 6:357
224. Vermue MH, Tramper J (1990) 5th European Congress on Biotechnology, vol 1, p 243
225. Kulla HG (1991) Chimia 45:81
226. Nikolova P, Ward OP (1992) J Ind Microbiol 10:169

227. Chang HN, Lee Y-H, Lee CY (1992) Ann N Y Acad Sci 15:643
228. Oda S, Ohta H (1992) Biosci Biotech Biochem 56:2041
229. Nikolova P, Ward OP (1992) In: Tramper J, Vermue MH, Beeftink HH, von Stockar U (eds) Biocatalysis in non-conventional media. Elsevier, Amsterdam, p 675
230. Harrop AJ, Woodley JM, Lilly MD (1992) Enzyme Microb Technol 14:725
231. Takeda K, Asou T, Matsuda A, Kimura K, Okamura K, Okamoto R, Sasaki J, Adachi T, Omura S (1994) J Ferment Bioeng 75:380
232. Prichanont S, Stuckey DC, Leak DJ (1994) In: Pyle DL (ed) Separations for biotechnology 3. SCI, London, p 392
233. Wubbolts MG, Hoven J, Melgert B, Witholt B (1994) Enzyme Microb Technol 16:887
234. Green KD, Gill IS, Khan JA, Vulfson EN (1996) Biotechnol Bioeng 49:535
235. Glöckler R, Roduit J-P (1996) Chimia 50:413
236. Kataoka M, Hirakata M, Sakamoto K, Yamada H, Shimizu S (1996) Enzyme Microb Technol 19:307
237. Lynch RM, Woodley JM, Lilly MD (1997) J Biotechnol 58:167
238. Vicenzi JT, Zmijewski MJ, Reinhard MR, Landen BE, Muth WL, Marler PG (1997) Enzyme Microb Technol 20:494
239. Molinari F, Bertolini C, Aragozzini F (1998) Biocat Biotransform 16:87
240. Kanda T, Miyata N, Fukui T, Kawamoto T, Tanaka A (1998) Appl Microbiol Biotechnol 49:377
241. Molinari F, Occhiato EG, Aragozzini F, Guarna A (1998) Tetrahedron Asymmetry 9:1389
242. Choi WJ, Choi CY, De Bont JAM, Weijers CAGM (1999) Appl Microbiol Biotechnol 53:7
243. Doig SD, Boam AT, Livingston AG, Stuckey DC (1999) Biotechnol Bioeng 63:601
244. Jaquet A, Quan L, Marison IW, von Stockar U (1999) J Biotechnol 68:185
245. Held M, Schmid A, Kohler HPE, Suske W, Witholt B, Wubbolts MG (1999) Biotechnol Bioeng 62:641
246. Omata T, Tanaka A, Fukui S (1980) J Ferment Technol 58:339
247. Saunders RP, Hardman R, Cheetham PSJ (1985) Biotechnol Bioeng 27:825
248. Kaul R, Mattiasson B (1986) Appl Microbiol Biotechnol 24:259
249. Ceen EG, Herrmann JPR, Dunnill P (1987) Enzyme Microb Technol 9:365
250. Steinert HJ, Vorlop KD, Klein J (1987) Biocatalysis in organic media. Elsevier, Wageningen, p 51
251. Boeren S, Laane C (1987) Biotechnol Bioeng 29:305
252. Bar R (1989) TIBTECH 7:2
253. Hocknull MD, Lilly MD (1990) Appl Microbiol Biotechnol 33:148
254. Silbiger E, Freeman A (1991) Enzyme Microb Technol 13:869
255. Ahmad S, Johri BN (1992) Indian J Microbiol 32:489
256. Pinheiro HM, Cabral JMS (1992) Enzyme Microb Technol 14:619
257. Lee CY, Chen CD, Liu WH (1993) Appl Microbiol Biotechnol 38:447
258. Houng JY, Chiang WP, Chen KC (1994) Enzyme Microb Technol 16:485
259. Dias ACP, Cabral JMS, Pinheiro HM (1994) Enzyme Microb Technol 16:708
260. Aono R, Doukyu N, Kobayashi H, Harushi N, Horikoshi K (1994) Appl Environ Microbiol 60:2518
261. Fernandes P, Cabral JMS, Pinheiro HM (1995) Enzyme Microb Technol 17:163
262. Liu WH, Horng WC, Tsai MS (1996) Enzyme Microb Technol 18:184
263. Evans PJ, Wang HY (1984) Appl Microbiol Biotechnol 47:1323
264. Dykstra KH, Li X-M, Wang HY (1987) Biotechnol Bioeng 32:356
265. Frenz J, Largeau C, Casadevall E, Kollerup F, Daugulis AJ (1989) Biotechnol Bioeng 34:755
266. Wang HY, Palanki S, Hyatt GS (1989) Appl Microbiol Biotechnol 30:115
267. Barenschee T, Scheper T, Schügerl K (1992) J Biotechnol 26:143
268. Yang WY, Lin CD, Chu IM, Lee C-J (1994) Biotechnol Bioeng 43:439
269. Paquet V, Myint M, Roque C, Soucaille P (1994) Biotechnol Bioeng 44:445
270. Kuboi R, Maruki T, Tanaka H, Komasawa I (1994) J Ferment Bioeng 78:431
271. Muller U, Merrettigbruns U, Hollmann D (1995) Chem Ing Tech 67:783
272. Hollmann D, Switalski J, Geipel S, Onken U (1995) J Ferment Bioeng 79:594

273. Kim CH, Kim SW, Hong SI (1999) Process Biochem 35:485
274. Persson I, Tjerneld F, Hahn-Hägerdal B (1984) Enzyme Microb Technol 6:415
275. Andersson E, Johansson A-C, Hahn-Hägerdal B (1985) Enzyme Microb Technol 7:333
276. Kantelinen A, Poutanen K, Linko M, Markkanen P (1987) Ann N Y Acad Sci 501:229
277. Andersson E, Hahn-Hägerdal B (1988) Appl Microbiol Biotechnol 29:329
278. Nguyen AL, Grothe S, Luong JHT (1988) Appl Microbiol Biotechnol 27:341
279. Park KM, Wang NS (1991) Biotechnol Prog 7:439
280. Persson I, Stalbrand H, Tjerneld F, Hahn-Hagerdal B (1992) Appl Biochem Biotechnol 27:27
281. Nilsson M, Vijayakumar AR, Holst O, Schornack C, Hakanson H, Mattiasson B (1994) J Ferment Bioeng 78:356
282. Chen JP, Lee MS (1994) Biotechnol Lett 8:783
283. Kuboi R, Umakoshi H, Komasawa I (1995) Biotechnol Prog 11:202
284. Umakoshi H, Yano K, Kuboi R, Komasawa I (1996) Biotechnol Prog 12:51
285. Illieva MP, Bakalova A, Mihneva M, Pavlov A, Dolapchiev L (1996) Biotechnol Bioeng 51:488
286. Kulkarni N, Vaidya A, Rao M (1999) Biochem Biophys Res Commun 255:274
287. Ramachandra Rao S, Ravishankar GA (1999) Process Biochem 35:341
288. Robins RJ, Rhodes MJC (1986) Appl Microbiol Biotechnol 24:35
289. Payne GF, Payne NN, Shuler ML, Asada M (1988) Biotechnol Lett 10:187
290. Asada M, Shuler ML (1989) Appl Microbiol Biotechnol 30:475
291. Buitelaar RM, Leenen EJTM, Tramper J (1992) Biocat 6:73
292. Corry JP, Reed WL, Curtis WR (1993) Biotechnol Bioeng 42:503
293. Kurata H, Kawai A, Seki M, Furusaki S (1994) J Ferment Bioeng 78:117
294. Byun SY, Pedersen H (1994) Biotechnol Bioeng 44:14
295. Bassetti L, Pijnenburg J, Tramper J (1996) Biotechnol Lett 18:377
296. Collins-Pavao M, Chin CK, H. P (1996) J Biotechnol 49:95
297. Hong HJ, Lee JE, Ahn JE, Kim DI (1998) J Microbiol Biotechnol 8:478
298. Mitsuda S, Matsuda Y, Kobayashi N, Suzuki A, Itagaki Y, Kumazawa E, Highashio K, Kawanishi G (1991) Bioprocess Eng 7:137
299. Zijlstra GM, Degooijer CD, Vanderpol LA, Tramper J (1996) Enzyme Microb Technol 19:2
300. Nayve FRPJ, Motoki M, Matsumura M, Kataoaka H (1991) Cytotechnol 6:121
301. Schneider M, Marison IW, von Stockar U (1994) Enzyme Microb Technol 16:957
302. Chang YHD, Grodzinsky AJ, Wang DIC (1995) Biotechnol Bioeng 47:308

Received: March 2002

Author Index Volumes 51–80

Author Index Volumes 1–50 see Volume 50

Subject Index

MIX
Papier aus verantwortungsvollen Quellen
Paper from responsible sources
FSC® C105338

If you have any concerns about our products,
you can contact us on
ProductSafety@springernature.com

In case Publisher is established outside the EU,
the EU authorized representative is:
Springer Nature Customer Service Center GmbH
Europaplatz 3, 69115 Heidelberg, Germany

Printed by Libri Plureos GmbH
in Hamburg, Germany